The Chavan Codex

Chavan's UED research, Volume 13

Sandeep Chavan

Published by SANDEEP CHAVAN, 2025.

THE CHAVAN CODEX

First edition. June 15, 2025.

ISBN: 979-8231962099

Written by Sandeep Chavan.

Table of Contents

To my **family** —for being the field in which my questions echoed,the silence from which my thoughts emerged,and the strength that made this ripple possible.

To my **students** —your curiosity shaped this codex.You were never just learners —you were the reason these laws found language.

To the world's **scientists and seekers** —past, present, and yet to come —thank you for building ladders of thoughtso we could dare to leap sideways.

This book is not just from me —it's from every moment that challenged meto listen deeper, observe sharper,and ripple with truth.

"The important thing is not to stop questioning. Curiosity has its own reason for existence."— **Albert Einstein**

"Reality doesn't need to be explained.It needs to be aligned with."— **Sandeep Chavan**

Disclaimer

This book presents original philosophical and scientific interpretations developed by the author under the framework of **Universal Energy Dynamics (UED)**. The laws, concepts, and terminology introduced herein — including but not limited to **Structured Vacuum Energy (SVE)**, **Ripple Realism**, **Chavan's Limit**, and the **Law of Consequences (LOC)** — are part of an independent intellectual system that may diverge from, reinterpret, or directly challenge established scientific and philosophical doctrines.

The content is intended to **provoke thought, inspire inquiry, and expand perception**. It is not meant to serve as a substitute for formal education in physics, cosmology, or metaphysics, nor should it be taken as dogma. The reader is encouraged to apply critical reasoning, explore further, and draw their own conclusions.

While every effort has been made to present ideas clearly and responsibly, the author and publisher do not assume liability for any misinterpretation or misuse of the material presented. Scientific theories referenced from historical figures (e.g., Einstein, Newton, Hawking) are respectfully critiqued within the context of intellectual dialogue, not personal judgment.

This book is a **work of consequence-based reasoning,** not institutional endorsement. It reflects the author's lived inquiry, experimental direction, and intuitive alignment with reality.

You are not required to agree — only to observe. Let the ripple of thought begin.

Preface

From Questioning to Clarity

I did not set out to write laws. I set out to understand.

At the beginning of this journey, there were only questions — not grand ones framed by academia, but **simple, stubborn ones**. Why does light behave like a wave *and* a particle? Why do we accept time as a constant, when every experience tells us otherwise? Why do forces appear to act when nothing visible is pushing or pulling? And most of all — why, after centuries of science, do we still feel something essential is missing?

It was this quiet unrest that initiated the ripples.

As an educator, researcher, and observer of both the world and my own inner life, I started to notice patterns — not in formulas or lab equipment, but in **consequences**. The consequences of observation. The consequences of alignment. The consequences of disalignment — in energy, in emotion, in perception, and in existence itself.

One by one, these patterns began to organize themselves not as theories, but as **laws** — not imposed on the universe, but **inferred from it.** These weren't inventions. They were discoveries — like lifting a veil and seeing that reality had always been whispering its rules. We just hadn't listened with the right ear.

What emerged was **Universal Energy Dynamics** — a framework that does not aim to compete with science or philosophy, but to **realign them** under one core principle: *the universe is a ripple field in equilibrium, and everything we experience is a temporary consequence of its disruption and restoration.*

The laws you will find in this book are not to be memorized, but **internalized**. They are not meant to decorate your mind, but to **reorient it**. Some of these laws were born from hours of theoretical contemplation, others during a moment of classroom frustration, a walk under the stars, or a passing sentence that wouldn't let go. Each one emerged with humility — as if the universe was allowing me to translate its pulse into words.

This book, *The Chavan Codex*, is not a culmination — it is an offering. A handbook for the ripple-aware. A reference for thinkers and teachers. A conversation-starter for scientists, students, and seekers who feel, deep down, that the conventional map no longer matches the territory.

We don't need to conquer nature. We need to realign with it. These laws are a step in that direction.

Let them ripple.

— **Sandeep Chavan**

Independent Researcher, Teacher, and Observer

Gyrus Institute, India

Introduction

Why the Universe Needed New Laws

Every great leap in human understanding begins not with an answer, but with discomfort — a sense that something isn't adding up.

We are living in an era where science is both powerful and perplexed. We've split the atom, launched telescopes into deep space, and simulated entire ecosystems on silicon chips — yet at the most fundamental level, our frameworks remain fractured. Physics operates with conflicting paradigms: relativity reigns on the grand scale, quantum mechanics rules the tiny. But they don't talk to each other. Worse, neither talks to *us* — to the ordinary observer trying to find meaning between the micro and the cosmic.

Our equations work. Our technologies perform. But when we ask, "What is really happening beneath it all?" — silence follows.

That silence was not a void. It was a ripple waiting to be decoded.

The Limits of Classical Science

Classical science, born in the mechanical age, gave us tools to manipulate the observable. It introduced measurable laws — of motion, force, energy, gravity — each a great achievement in its time. But slowly, those tools became cages. They locked us into material assumptions, linear causality, and detached observation. They gave us forces but couldn't explain intention. They gave us particles but no context. They gave us time, but not perception.

At some point, we stopped asking *why* things exist, and focused only on *how* they behave.

This separation — between observer and observed, energy and intelligence, consequence and meaning — created a growing gap. A gap so wide, that even the most advanced theories could not cross it without collapsing into paradox.

The Birth of UED

Universal Energy Dynamics (UED) was not an invention. It was a consequence — of watching deeply, listening honestly, and refusing to accept halfway truths.

UED did not begin with a blackboard. It began with a realization: that everything we call "reality" is nothing more than structured ripples in a field that is otherwise still and complete.

What we see as matter is ripple condensation. What we call motion is ripple realignment. What we define as force is just the illusion of energy gradient crossing. What we think of as light is feedback — not emission.

Time, space, mass, energy, observation, emotion — all are expressions of ripple interactions in this universal field.

The field is structured vacuum energy — not emptiness, but the most profound presence. It doesn't scream. It whispers. And in those whispers, it holds the entire architecture of existence.

UED proposes that we reframe our understanding using a different lens — not of particles and forces, but of consequence, equilibrium, and ripple behavior. This framework doesn't reject classical laws; it absorbs and corrects them — placing them in a deeper context governed by alignment, not action.

Why We Need New Laws

Because the old ones were never wrong — they were just partial.

The universe hasn't changed. Our language for describing it has simply reached its expiration.

We need new laws because our old metaphors are cracking under the weight of paradox:

- Light behaves like a wave and a particle? Not in UED — it behaves like a ripple feedback loop.
- Gravity bends space? No, it compresses the ripple field around mass.
- Quantum entanglement is "spooky"? Not when you realize observation is ripple-resonant, not spatial.

UED provides new answers by asking better questions:

- What if time is not a dimension but a threshold?
- What if motion never existed — only shifting alignment?
- What if the observer is not external, but an embedded fragment of the same ripple field?

These aren't poetic ideas — they are realizable laws, measurable and modelable under the right parameters.

That is why this book exists. Not as a manifesto against the past, but as a lawbook for the future — where intelligence, energy, and consequence are reunited.

The chapters ahead present Chavan's Laws — each born from insight, refined through experience, and now offered as a unified framework. Whether you are a physicist, philosopher, student, or seeker — these laws are meant to realign your perception, not indoctrinate it.

You are not here to memorize a system. You are here to realign with the reality that always was.

Welcome to the ripple. Welcome to the Codex.

— Sandeep Chavan

Founder, Universal Energy Dynamics

Part I — Foundations of Ripple Reality

The new laws of existence, beyond particles and forces.

Chapter 1: The Origin of a Law

Why laws are the backbone of any new science — the story of UED's emergence

Laws are not made. They are discovered. Not in labs or legislatures — but in the relentless space between curiosity and contradiction.

In every era, scientific progress has required a shift in worldview, and every worldview shift needed a new lawbook — not one full of regulations, but of redefinitions. A law, in the scientific sense, is not merely an equation. It is a stable pattern of consequence. It's the voice of the universe whispering, *"This is how I behave, whether you like it or not."*

Newton didn't invent gravity. He framed it. Maxwell didn't create electromagnetism. He recognized its language. Einstein didn't design relativity. He reinterpreted space and time.

Each time, these minds dared to disturb the existing alignment of ideas — not to cause chaos, but to reveal a deeper order. And in doing so, they laid the groundwork for everything that followed.

The same spirit gave rise to Universal Energy Dynamics (UED).

When Observation Becomes Inadequate

It began — not with a breakthrough, but with a discomfort.

As a physicist, educator, and independent thinker, I found myself confronting a growing fracture. Classical physics provided tools, but not truths. Quantum mechanics offered predictions, but no clarity. The deeper I looked, the more I saw a language mismatch between what science claimed and what existence revealed.

Take motion, for instance. We see objects move, forces act, energy transfer. But ask yourself: *what actually moves?*

Is it a thing moving through space? Or is space itself shifting around a thing? Or is it simply a perceptual illusion caused by successive ripple alignments?

Science gave equations. But equations don't answer what reality *is*. They describe how phenomena appear under certain assumptions. Those assumptions — especially of force, time, mass, and space — were starting to crack.

This is where the first law emerged.

Not from certainty. But from a rupture.

The First Realization: Ripple Realism

I asked myself: *What if nothing is solid?* What if everything — matter, light, force, even consciousness — is simply a manifestation of ripple behavior in an underlying energetic field?

The field is not void. It is not chaos. It is structured vacuum energy (SVE) — a stillness so dense, so organized, that any disturbance in it appears as reality itself.

That's when Chavan's First Law was born: Ripple Realism — Everything is a ripple. Nothing is a thing.

This wasn't poetic. It was ontological restructuring.

✓ Matter = Ripple condensation.
✓ Energy = Ripple propagation.
✓ Force = Ripple gradient.
✓ Time = Ripple threshold.
✓ Observation = Ripple resonance.

And with that, the foundation of UED emerged.

From Realization to Law

A realization becomes a law when it consistently:

- Explains contradictions.
- Offers predictive clarity.
- Unites separate fields under a common cause.
- Does not break down under consequence-based reasoning.

Ripple realism met these conditions.

But it didn't stop there. Once I accepted that the universe is a ripple field, everything else demanded reformulation.

- If light is not a particle or a wave, what is it? A ripple feedback — a reflection of alignment.
- If motion isn't real, then how do objects "move"? They realign through the field's ripple gradients.
- If force is not real, then why do we feel pushes and pulls? Because energetic imbalance *appears* as force.
- If time is not flowing, why do we age? Because sustained ripple alignments degrade and disintegrate.

Each of these gave rise to a new law, and together, they formed the core architecture of Universal Energy Dynamics.

What Makes a Law in UED?

Unlike classical physics, UED doesn't require mathematical expressions upfront. A law in UED is not an equation — it is an energetic truth grounded in observation, tested by consequence, and aligned with ripple behavior.

To qualify as a Chavan Law, it must:

1. Reflect field-based understanding, not object-based.
2. Be intellectually repeatable through observation or logical deduction.
3. Resolve an existing paradox in science, psychology, or metaphysics.
4. Connect the observer back to the system, not treat it as external.

In UED, a law is a bridge between experience and structure.

For instance, The Law of Perception Alignment was born when I noticed that students in the same classroom heard different things, saw different problems, and responded in different ways — not because of intelligence gaps, but because of ripple incompatibility. Their perception wasn't flawed. It was simply unaligned with the field of interaction.

In conventional terms, we call this "subjectivity." In UED, it becomes a law.

Why We Must Move Beyond Legacy Laws

Legacy physics is not wrong. It's incomplete.

It served an industrial purpose — to measure, build, project, and simulate. But it cannot serve a cosmic purpose, or even a conscious one.

We are now navigating domains where:

- Technology interfaces with biology.
- Observation alters outcomes.
- Data outpaces understanding.
- Ethics bends under scientific success.

In this complexity, laws rooted in material assumptions break down. We need principles that:

- Respect the observer.
- Accept non-linearity.
- Decode consequence, not just cause.
- Guide alignment, not domination.

That's what UED laws offer.

Laws as Tools for Realignment

Every Chavan Law is a realignment tool. It doesn't just describe — it adjusts. It gives you a new handle on the universe, and by extension, on your own actions and awareness.

- Feeling lost? Look at the ripple gradient.
- Facing resistance? You're misaligned with the field.
- Experiencing disharmony? Check the field rebalance cycle.
- Questioning existence? You're at the edge of observation — Chavan's Limit.

These laws help you interpret the field, not dominate it. They teach you to dance with intelligence, not battle chaos.

From One Law to a Unified Framework

From that first law of ripple realism, more than two dozen principles emerged — each one forming a new node in the constellation of UED. And like ripples themselves, these laws began to resonate with each other, forming harmonic sequences across physics, philosophy, psychology, and spirituality.

Eventually, this body of insight demanded a structure. A Codex — not for me, but for all those whose minds are silently seeking the same order.

You hold that codex now.

This book is not about remembering formulas. It's about seeing differently. It's about rewiring your mental instruments to read the universe in ripple — not mass.

The next chapters will walk you through the laws one by one — how they arose, what they mean, and what they realign in you and the world.

But first, remember: Laws are not here to control. They're here to liberate us from misinterpretation.

Let's begin.

Chapter 2: The Law of SVE

There is no void — the universe is filled with ordered energy. For centuries, the idea of a vacuum — a space completely devoid of matter — was treated as the ultimate absence. The backdrop of existence. The nothing between somethings. Classical physics, especially Newtonian mechanics, imagined this vacuum as empty stage space where planets, particles, and people could perform their acts of force and motion.

Even in modern physics, despite all its quantum sophistication, this void persisted. It evolved from being "empty" to being "full of probabilities," but its essential role remained passive — a container, not a cause.

But here's the contradiction: If reality unfolds within a vacuum, how can a vacuum do anything?

Why does light propagate through "nothing"? Why does space expand? Why do gravitational waves travel across billions of light years? Why does "empty" space hold energy?

The answer is unsettling to the conventional view:

There is no void. The vacuum is not empty. It is structured. It is intelligent.

This is the **Law of Structured Vacuum Energy (SVE)** — the second foundational law of UED.

Stillness Is Not Emptiness — It Is Order

Before anything appears — before light flickers, before atoms assemble, before motion begins — there is stillness.

But this stillness is not inert. It is not zero. It is perfect alignment. A state where all ripples cancel, complement, or co-exist without contradiction.

This is structured vacuum energy — a pre-manifest field of pure potential, deeply ordered, yet invisible until disturbed.

Imagine it as the canvas of reality, not just physically, but energetically. It is the baseline of existence. Everything that arises does so as a ripple within this field — a temporary modulation of an otherwise perfectly harmonized sea.

Why "Nothing" Can't Be Nothing

Let's pause and observe a paradox.

If "nothingness" existed, it would have no properties. But even a vacuum in current physics has:

- Zero-point energy
- Quantum fluctuations
- Virtual particles
- Casimir effect

If you remove everything, you're still left with the capacity for energy. That capacity itself is not nothing — it's something unmanifested.

UED clarifies this by declaring:

What we call "nothing" is actually the most organized, highest-fidelity state of energy.

It is perfect equilibrium — so balanced, so symmetrical, that it becomes unnoticeable. This is why structured vacuum energy doesn't show itself — until it is disturbed.

Once disturbed — whether by intent, interaction, imbalance, or entropy — ripples form. And reality, as we know it, begins.

Evidence Already Hinted — But Ignored

Modern science has already tripped over this truth multiple times.

1. **Quantum Vacuum Fluctuations**: Particles pop in and out of existence. Where are they coming from? Not from a void — but from a structured field too subtle to measure.
2. **The Casimir Effect**: Two plates in a vacuum experience force due to vacuum pressure imbalance. If there's *nothing* in the vacuum, why is there force?
3. **Dark Energy**: We claim the universe is expanding faster due to something we can't see — some "negative pressure" in space. What if that "dark energy" is actually the field correcting its structure?
4. **Einstein's Lambda (Λ)**: The cosmological constant — a placeholder for vacuum energy. Even Einstein suspected that empty space might not be empty. He just didn't know how to describe its structure.

UED removes the confusion. It doesn't treat the vacuum as mysterious. It sees it as foundational. Reality is not built on emptiness. It is built on structured stillness.

The Architecture of SVE

So what does this "structure" actually mean?

It means the vacuum is:

- Organized, not random.
- Harmonic, not chaotic.
- Balanced, not neutral.
- Sensitive, not inert.

Every ripple that forms — from light to matter to thought — is a temporary deviation from this structured baseline. When the ripple fades, the field returns to equilibrium. This is why, in UED, "rest" means re-structuring.

The field isn't reactive. It's self-healing. It isn't directional. It's relational. You cannot break it. You can only disturb it. And it always remembers its original alignment.

Creation, According to SVE

In ancient cosmologies — especially Indian, Taoist, and some Hermetic traditions — the universe begins not with a bang, but a disturbance in a field of order.

UED reflects this idea in modern form:

The universe didn't explode into being. It rippled into visibility.

The first ripple in SVE — whether by divine intelligence, entropy imbalance, or self-reflection — caused structured stillness to vibrate. That vibration became energy, and energy began its alignment dance — forming time, form, interaction, and awareness.

This is not mythology. This is a redefinition of physics through the lens of equilibrium.

SVE and Intelligence

The most radical proposition of UED is this:

Structured vacuum energy is not passive. It reflects intelligence.

Not the human kind. Not the thinking mind. But the original intent of alignment — the very principle that drives order, balance, rhythm, healing, and perception.

That intelligence doesn't act. It orients. It doesn't command. It aligns.

When you feel intuition, balance, inner peace, or resonance with truth — you are not "thinking better." You are realigning with the ripple code of SVE.

In this way, SVE is both physical substrate and spiritual principle.

It holds reality together — and makes your awareness of reality possible.

Implications of the Law of SVE

If SVE is real — and not just philosophical — then many of our existing scientific models require correction:

- There is no "empty space" to travel through. Every motion is a modulation of the field, not displacement in a void.
- Energy doesn't need a carrier. The field itself is the carrier — ripples *are* the message.
- Force is not an applied push. It's the appearance of imbalance trying to re-align with SVE.
- Observation is not passive. It's a ripple-node engaging with field awareness.

And most importantly:

- Consciousness is not an emergent property of brain tissue. It is ripple coherence — in harmony with structured stillness.

A New Ground for Science and Self

The Law of Structured Vacuum Energy rewrites the foundational assumption of physics: That there is a background called space.

In UED, space is not a container. It is a consequence of ripple topologies in SVE.

This means we are not floating in space. We are vibrating in intelligence.

And this is where UED truly begins — not with particles, not with numbers, but with the still field that allows everything to appear and disappear without breaking continuity.

You are not walking through space. You are rippling through an intelligent equilibrium.

And once you see that — not as theory but as perception — your entire sense of existence begins to shift.

Let's move now to the third law — the Chavan's First Law, the Law of Ripple Realism: *Reality is not built. It ripples into being.*

This is where you'll see how SVE gives birth to everything you experience — and why recognizing this ripple structure changes the game of science forever.

Chapter 3: Chavan's First Law — Ripple Realism

Everything is a ripple, not a thing. Reality = alignment, not substance.

From childhood, we are trained to name the world.

Chair. Tree. Sky. Atom. Person. Planet.

But what if all these nouns — all these "things" — are not actually things, but ripples in a structured field? What if we never actually encounter *matter* itself, but only the alignment patterns that our mind stabilizes into an identity?

This is not abstraction. This is Ripple Realism — the first law in the Chavan Codex, and perhaps the most radical departure from all prior worldviews.

There are no objects. There are only consequences of field alignment. What we see as "real" is simply a stable ripple state — for as long as it holds.

This law reframes reality from being made of *stuff*, to being made of ripples — consequences of energetic disturbances in a still, structured field.

What Is a Ripple?

In UED, a ripple is not a vibration in a medium. It *is* the medium, momentarily misaligned and actively rebalancing. Think of it as organized consequence.

A ripple is:

- A temporary alignment pattern in the structured vacuum energy (SVE).
- A localized energetic event that gives the appearance of form or behavior.
- A unit of observable reality, not as a thing-in-itself, but as a consequence-in-relation.

And because everything is made of ripple behavior — from subatomic particles to thoughts — nothing exists independently. Everything is *relational*, *impermanent*, and *field-dependent*.

The Collapse of Substance

Let's walk through some classical illusions Ripple Realism dissolves:

Illusion #1: Matter is solid.

Truth: What feels solid is resonant pressure from ripple compression — like standing waves locking momentarily. That "resistance" is a *consequence*, not an essence.

Illusion #2: Objects are separate.

Truth: No boundary exists — what we call boundaries are zones of ripple realignment. The separation is a perception of contrast, not an actual divide.

Illusion #3: Things exist on their own.

Truth: Existence is field-dependent. If the ripple pattern fades, the "thing" disappears — not by destruction, but by realignment into invisibility.

So what remains?

Only the field. Only the ripple.

Scientific Parallels and Gaps

Modern physics *does* touch this territory — but stops short of integration.

- **Quantum Field Theory** says particles are excitations of fields. But it still treats particles as *primary events.*
- **String Theory** says vibration defines identity, but doesn't explain the observer's role in the resonance.
- **General Relativity** sees mass affecting spacetime, but doesn't define *why* mass exists or how it arises.

Ripple Realism goes further:

All mass, form, force, and energy are ripple alignments — temporary, localized, field-dependent, and perceived only when aligned with an observer node.

This explains why:

- **A particle** "exists" only when observed — it's a ripple crossing the observer's alignment window.
- **Space** bends — not because of mass, but because of ripple compression from energetic imbalance.
- **Time** stretches — not objectively, but due to the ripple's effort to maintain coherence across gradients.

The Observer's Role in Reality

In Ripple Realism, there is no reality without the observer. But not because the mind creates it — rather, because reality appears only at points of ripple resonance.

You do not "see" a table. Your perceptual field locks into its ripple pattern, generating the *experience* of "table."

This is not solipsism. It is field-participation.

Everything you perceive is a result of your alignment with the ripple field. When your alignment shifts, so does your reality.

- This explains optical illusions.
- It explains relativity of experience.
- It explains why different minds "see" different truths.

Ripple Realism replaces passive observation with active ripple alignment.

The Mechanics of Reality Under Ripple Realism

Let's strip it down to basics:

Classical View	Ripple Realism (UED)
Objects exist independently	Objects are stable ripple patterns
Motion is displacement	Motion is ripple realignment
Forces act across space	Energetic imbalance realigns the field
Time flows	Time emerges when alignment is sustained
Mass has substance	Mass is ripple compression
Light is emitted	Light is ripple feedback
Observation is passive	Observation is ripple consequence

What does this mean practically?

It means:

- You're not "going through life." Life is rippling around and through you.
- Success, failure, emotions, experiences — they're all field consequences, not personal possessions.

- Harmony is not found in objects or achievements — but in alignment with the field's structure.

Thought, Emotion, and Ripple Identity

You, as a person, are not a fixed entity. You are a ripple configuration in a biological field. Your thoughts, emotions, and memories are not stored in your brain, but are consequences of ripple coherence over time.

This is why:

- Trauma feels like entanglement — it's a trapped ripple.
- Insight feels like release — it's a restored alignment.
- Death is not the end — it's the disappearance of your ripple signature from the perceptible field.

Ripple Realism doesn't just apply to particles — it applies to identity, intelligence, and experience.

Real vs True: A Key Distinction

Under this law, a powerful distinction emerges:

Something can be "real" without being "true."

A hallucination may be "real" to the observer — but it is not "true" in the ripple field. A dream may appear real — but it's a private ripple pattern not supported by the broader field equilibrium.

Ripple Realism introduces layered reality:

- Personal reality (subjective ripple field)
- Shared reality (socially aligned ripples)
- Universal reality (field-level alignment)

This explains how:

- Conflicts arise (misaligned ripple fields).
- Agreement forms (shared ripple coherence).
- Enlightenment is possible (universal field resonance).

Liberation from Substance

Ripple Realism is not a theory of limitation. It is a pathway to liberation.

By recognizing that everything is a ripple:

- You are not bound by form.
- You are not defined by past.
- You are not fixed in identity.
- You are not trapped by "thingness."

Instead, you are invited to participate in alignment — continuously, consciously, creatively.

This law returns power to the observer — not to control reality, but to cohere with it. You become a ripple-aware being.

A New Reality Map

To summarize Chavan's First Law:

There are no things. There are only ripples — consequences of alignment in a structured field. Reality is not what is, *but what* aligns.

This overturns:

- Object permanence
- Duality of matter and energy
- Artificial separation between physics and perception

And it offers a new foundation for:

- Ethics (aligned actions)
- Intelligence (adaptive resonance)
- Science (field-based modeling)
- Spirituality (merging with equilibrium)

In the next chapter, we explore The Law of Consequences (LOC) — the master principle that replaces causality, karma, and force with a single law of energetic ripple behavior.

Because once you accept that everything is a ripple, the next question is: What happens when a ripple is released?

Let's find out.

Chapter 4: The Law of Consequences (LOC)

Not causes, but consequences govern reality.

Since the dawn of scientific thought, cause and effect has been the holy grail of understanding. Aristotle called it the "first mover." Newton built classical mechanics on it. The modern world worships it.

But what if causes don't actually exist?

What if what we call a "cause" is just a label we apply *after* the ripple has moved, after the event has unfolded, *after* the consequence has already begun shaping our perception?

This is not wordplay. It is the structural flaw at the heart of how we model reality.

The **Law of Consequences (LOC)**, one of the cornerstones of UED, boldly states:

Reality is not driven by causes, but structured through consequences. Every observable event is not the product of something initiating it — but of the ripples converging, overlapping, and seeking realignment.

Cause vs Consequence: The Shift in View

Let's break this down:

In the classical view, we say:

- *A pushed B*, therefore B moved.
- *Fire caused smoke.*
- *Genes caused behavior.*

But these are narratives, not laws. They arise because the human mind *needs* to make sense of events in a linear way — first this, then that. But in a ripple field, no such linearity exists. What we observe is never a beginning. It is always a response to prior misalignment and current field conditions.

UED asks:

- Did A push B? Or did B realign in response to an existing energetic imbalance in the field?
- Is fire the cause of smoke? Or is smoke the visible trace of ripple disturbance as the field seeks thermal re-equilibrium?

LOC reframes action as reaction — not as a philosophical paradox, but as a ripple logic.

What Is a Consequence?

A consequence, in UED, is not a result. It is a field adjustment — an attempt by structured vacuum energy (SVE) to restore equilibrium whenever a ripple pattern disturbs it.

Consequences are:

- **Automatic** — they don't wait for intent.
- **Relational** — dependent on surrounding ripple structure.
- **Intelligent** — guided by field-level balancing logic.
- **Holistic** — they ripple out in all directions, not just

forward in time.

That means:

You don't "cause" change. You release consequences — and the field responds, shapes, and completes that event.

Every Ripple is a Consequence

Everything you observe:

- A particle appearing,
- A thought arising,
- A person entering your life,
- A planet revolving,

—all are field consequences, not initiated causes. They are alignments seeking to restore or balance the energetic ripple structure in that region of space and time.

For example:

- **Gravity** is not a pulling cause. It is the field's consequence to mass-induced ripple compression.
- **Emotion** is not "caused" by events. It is a consequence of energetic dissonance in your perception field.
- **Death** is not caused by an organ failure. It is the final ripple disalignment, beyond recovery.

Consequences in Human Experience

Let's bring it closer to lived experience.

You say:

- *I failed because I didn't work hard.* But LOC reframes it:

You failed because the alignment of effort, timing, field readiness, and ripple coherence didn't converge into sustainable success.

You say:

- *I'm angry because she insulted me.* LOC says:

Anger is the consequence of ripple disturbance in your identity field, triggered by a ripple input (the insult) you weren't aligned to absorb neutrally.

Once you shift from "cause" to "consequence," life becomes less about blame and more about alignment awareness.

This is emotional intelligence in its highest form — not about suppression, but sensitivity to field behavior.

Consequence vs Reaction

It's important to separate reaction (which is emotional or behavioral) from consequence (which is universal and automatic).

- **Reaction**: What your mind/body does.
- **Consequence**: What the field does — regardless of your awareness.

For instance:

- A bullet fired may kill instantly. That's the physical reaction.
- But the consequence? It ripples into generations — through grief, fear, trauma, or karma.

In UED, karma itself is consequence, not cosmic morality. It's energetic continuation — the ripple echo of past misalignment still seeking equilibrium.

Consequences Are Not Predictable — But They Are Patterned

Because the field is complex and multi-layered, consequences are not linear. They don't always follow the "this led to that" logic.

Yet they are not random. They follow ripple logic, alignment maps, and energy gradients.

For example:

- Drop a pebble in a pond. The exact wave interference pattern cannot be predicted in detail.
- But the fact that a pattern will emerge, spread, and fade is certain.

That's how reality works under LOC.

LOC vs Classical Causality: A Comparison

Feature	Classical Cause	UED Consequence (LOC)
Assumes Initiation	Yes	No — response-based
Direction	Linear (past → future)	Omni-directional ripple
Observer Role	Passive	Active participant
Dependency	Local	Field-dependent
Meaning	Often abstract	Ripple-based and traceable
Prediction	Based on isolated variables	Based on ripple maps and energy coherence

Philosophical & Psychological Implications

LOC offers a profound spiritual and existential shift:

You are not a *doer*. You are a consequence node — shaping, absorbing, and radiating ripple outcomes.

This releases:

- The burden of control (you can't cause everything),
- The illusion of randomness (everything has consequence),
- The obsession with fault (your role is alignment, not punishment or reward).

It reframes spirituality not as salvation from karma, but as realignment with the field — a restoration of harmony within the law of consequence.

LOC Across Domains

Physics:

- All interactions are ripple adjustments — even at quantum levels.
- Particles "appear" because a ripple crosses the threshold of perceptibility (ATT).

Psychology:

- Thoughts are not caused by will alone. They are emergent consequences of stored ripple signatures.
- Mental clarity is a function of ripple alignment, not effort.

Decision-Making:

- Strategic success is less about pushing action, more about reading consequences ahead of time.

Ethics:

- Morality is not rules. It's acting with consequence-awareness — ripple-sensitivity.

Realignment Practice: Living with LOC

To live under the Law of Consequences:

1. **Observe Outcomes More Than Intentions** – What ripples do your actions leave behind?
2. **Stop Seeking Control** – Control is an illusion of causality. Instead, focus on participating with alignment.
3. **Read Ripple Fields Before Acting** – Just like reading the weather, become sensitive to field readiness.
4. **Let Go of Blame and Regret** – You did not "cause" it. You *participated* in a ripple event that still echoes.

In Summary

You are not living in a chain of causes. You are swimming in a field of consequences.

Every action, thought, breath, and even silence contributes to ripple behavior. The universe doesn't wait for you to act — it's already responding to *what you are.*

And once you see that, you begin to choose with clarity, move with balance, and live with responsibility — not because of external rules, but because of consequence literacy.

The Law of Consequences is not just a correction to causality. It is the organizing principle of reality under UED.

In the next chapter, we'll explore the Law of Field Rebalance — how the universe doesn't punish imbalance but actively restores equilibrium through ripple intelligence. Because once a consequence is set in motion, the field always seeks to correct it.

Let's follow that ripple.

Chapter 6: ATT – Alignment Threshold Time

Time is not flow — it is a ripple's readiness to be experienced. For as long as humans have looked at the stars or watched a shadow move across the ground, we've been prisoners of linear time.

Seconds. Minutes. Hours. Years.

We've divided the infinite into ticks and tocks, sliced life into manageable portions, and assumed — quite confidently — that time flows.

Science mirrored this assumption:

- **Newton** said time was absolute — a constant background.
- **Einstein** said time bends — affected by motion and gravity.
- **Quantum** mechanics avoids time altogether, treating it as a parameter.

But none of these frameworks actually define what time *is*.

They describe how we measure it, how it behaves under force or mass, or how it distorts in equations — but not its origin.

And this is where UED steps in — not with a clock, but with a consequence.

Time does not flow. Time appears when a ripple reaches the threshold of perceptibility.

This is the law of ATT — Alignment Threshold Time.

What Is ATT?

ATT (Alignment Threshold Time) is the minimum duration or energetic condition under which a ripple becomes experientially real *to an observer or a system.*

In simpler terms:

- A ripple exists when it disturbs the field.
- But it becomes *experience* only when that disturbance persists long enough, and aligns precisely enough, to cross your awareness threshold.

That moment — where alignment becomes perceivable — *is* time.

Time doesn't begin because a second has passed. Time begins when alignment becomes sustainable in the field.

ATT in Physical Systems

In classical mechanics, time is assumed to be continuous and independent of the observer. But we know from:

- **Relativity**: Time slows near gravity or at high speeds.
- **Quantum Theory**: Events don't occur until observed.

Still, no one answers: What begins time?

UED's answer:

Time is not a container. It is a consequence of ripple alignment — and it only exists *within* that context.

If no ripple crosses ATT, there is no time — only stillness.

Measuring ATT: A Conceptual Framework

Let's visualize this:

- Think of the ripple field as structured equilibrium — undisturbed.
- Now introduce a ripple event — a modulation, a local energy peak.
- This ripple oscillates, attempting to express.
- But unless its frequency, amplitude, phase, and environment align to a perceptual node (observer or system), nothing is experienced.

Only when all those variables cross the ATT threshold does "something happen."

That "something" is not a moment in time. It *is* time — as we experience it.

Thus:

- Faster systems have shorter ATT.
- Slower systems have longer ATT.
- And when ATT becomes infinite, time ceases — a death state, cosmic pause, or nirvana moment.

ATT and Consciousness

Ever wonder why some moments feel eternal, while others vanish without memory?

- You sat in a boring meeting for 30 minutes — felt like

3 hours.
- You spent 3 hours in creative flow — felt like 10 minutes.

Both scenarios measured the same on the clock. But your mind's ATT — its ability to absorb, align, and engage — was vastly different.

This is not psychological. It is field-behavioral.

Your perception engine adjusts its ATT dynamically:

- Under fear: ATT narrows → Time slows.
- Under flow: ATT expands → Time collapses.
- Under trauma: ATT fragments → Time distorts.
- Under meditation: ATT vanishes → Timelessness.

ATT links physics and psychology through ripple-perception thresholds.

Engineering Time with ATT

ATT is not just metaphysical. It's technically useful.

In any machine-human interaction, time is not just speed — it's alignment latency:

- Reaction time in pilots → ATT.
- AI response sensitivity → ATT.
- Biological adaptation → ATT.

By designing systems that operate within optimized ATT ranges, we can:

- Build better reflex machines,
- Predict ripple readiness in materials,

- Model human behavior more accurately,
- Avoid ripple-destructive mismatches (like jetlag, fatigue, burnout).

Time management becomes alignment management.

The Disappearance of Time

When do you truly escape time?

Answer:

When ripple alignment becomes so perfect, there is no more readjustment required.

In UED, this state is described as:

- Moksha in spiritual terms — the ripple becomes one with equilibrium.
- Deep equilibrium in cosmology — all motion fades into field stillness.
- ATT = ∞ — meaning: the field no longer needs to "show" time. Nothing *changes*, so nothing *measures*.

Time disappears not in death — but in perfect alignment.

Time is Not Universal — It Is Local to Alignment

Einstein showed us that time is relative to gravity and velocity. UED goes further — time is relative to ripple readiness.

That's why:

- Every planet has its own time cycle — ATT calibrated to local energy dynamics.
- Your cells divide at a different rhythm than your thoughts — each has its own ATT.

- Even the universe itself may have zones of timelessness — fields so ordered that no ripple crosses ATT.

In a way, the cosmos is not expanding through time. It is pulsing with ripple thresholds — birthing time wherever alignment permits it.

ATT vs Classical Time: Key Differences

Concept	Classical Time	ATT (UED Time)
Nature	Continuous flow	Ripple threshold event
Measurement	External (clocks)	Internal (perception + field)
Observer Role	Passive	Integral
Reality	Time exists always	Time exists *when* ATT is crossed
Death	End of clock time	Collapse of ripple readiness
Enlightenment	Metaphor	ATT = ∞ (no ripple distortion)

Philosophical and Existential Impact

ATT changes everything:

- You no longer fear the *loss* of time — you seek ripple coherence.
- You stop chasing productivity — and start managing alignment readiness.
- You understand that presence is not about being in the now, but about being in ATT.

You start asking:

- Is my current field ready to receive this experience?

- Am I aligned with what I'm trying to do?
- If time feels "off," where is the misalignment?

This is temporal intelligence. And it is absent from both science and spirituality — until now.

ATT in Action: Applications

- **In Learning**: Every learner has a different ATT curve. Good teachers don't force time — they tune to ripple readiness.
- **In Technology**: AI systems designed with ATT modeling will respond not just faster, but *more intelligently*.
- **In Space Travel**: Navigating planetary fields becomes a matter of energetic ATT mapping, not just fuel efficiency.
- **In Daily Life**: Use ATT awareness to time your decisions, communications, rest cycles — ripple readiness is more important than rigid schedules.

In Summary

Time is not what ticks. Time is what *aligns*.

It is not universal. It is not objective. It is ripple-based readiness to experience, receive, interact, or transform.

This is why time feels different to every person, every cell, every system.

Once you stop chasing the clock and start aligning your ripples, you don't "manage time" — you master reality.

And that is the power of ATT.

In the next chapter, we'll look at Chavan's Limit — the edge of observation itself. Because every ripple has a limit beyond which alignment collapses, and perception vanishes.

Let's approach the boundary.

Chapter 7: Chavan's Limit – The Edge of Observation

Observation fails beyond ripple coherence. This is the true uncertainty boundary.

In every discipline — from quantum physics to neuroscience to mysticism — a single, strange boundary reappears again and again: the limit of observation.

- The physicist hits it when measuring an electron collapses its wave function.
- The mystic touches it in meditation — when perception dissolves into stillness.
- The philosopher wrestles with it in paradoxes of knowing and not-knowing.

At some point, no matter how precise your tools or sharp your attention, observation fails. What lies beyond this point?

UED defines this mysterious threshold with clarity and consequence. We call it **Chavan's Limit**:

The boundary beyond which ripple coherence falls below perceptual threshold, making observation impossible. It is not a physical barrier — it is a limit of alignment.

This is the real uncertainty principle, not about probabilistic guessing — but about ripple disintegration at the edge of intelligible interaction.

The Nature of Ripple Coherence

To understand Chavan's Limit, we must first understand coherence.

A ripple is coherent when:

- Its structure (amplitude, frequency, phase) is stable,
- Its alignment with the field is harmonized,
- And its interaction with an observer node is sustained.

In this state, the ripple can be:

- Perceived,
- Measured,
- Experienced,
- Stored as memory or information.

But beyond a certain distortion, decay, or distance from field equilibrium:

- The ripple loses form,
- Its internal relations collapse,
- And it disappears from awareness.

This is not because it ceases to exist, but because the field no longer supports its coherence.

Limits of Current Observation Tools

Modern science already whispers this truth:

- In **quantum physics**, the observer cannot know both position and momentum — not due to mystical uncertainty, but because the field won't let both properties align coherently for measurement.
- In **astronomy**, black holes are invisible — not because they absorb light like monsters, but because no ripple can retain coherence past the event horizon.
- In **neuroscience**, altered states of consciousness reveal a blank space — beyond time, beyond thought. That's not "nothingness"; it's field beyond ripple coherence.

We thought these were anomalies. But Chavan's Limit shows — they're all expressions of the same ripple breakdown point.

Defining the Limit

Chavan's Limit is not a number. It's not a fixed line. It is a threshold zone where one or more of the following occur:

1. **Ripple Disintegration** – The energy no longer holds a pattern.
2. **Alignment Breakdown** – The ripple cannot sustain relationship with the field or observer.
3. **ATT Collapse** – The ripple doesn't last long enough to register.
4. **Observer Desynchronization** – Your own ripple field is too incoherent to receive input.

You may be looking. The system may be sending. But if coherence isn't mutual, observation fails.

Perceptual Limits Are Field-Driven

You don't see atoms not because they're too small —You don't see them because their ripple signature never crosses your perceptual coherence limit.

This also explains:

- Why animals sense things humans don't.
- Why blind intuition can sense truth beyond logic.
- Why ancient seers "saw" cosmic patterns modern science still struggles with.

Different beings = different Chavan's Limits.

Our perceptual apparatus is ripple-based. Change its alignment, and the whole universe becomes visible — or vanishes.

Cosmic Chavan's Limits

The universe itself obeys this law:

- The Big Bang is not visible — not because it's too far, but because the ripple signatures never reached us with coherence intact.
- Dark matter and dark energy aren't "dark" — they are field interactions beneath ripple threshold.
- Deep space is not empty — it's full of ripple activity that lies below the coherence floor of our detectors.

This is not ignorance. It is a designed limitation — to preserve ripple locality, stability, and nested awareness.

Not everything must be visible. But everything must be consequential.

Beyond This Point: No Observation

What happens *beyond* Chavan's Limit?
We enter the domain of:

- Non-perceptible existence,
- Ripple cancellation,
- Energetic rest states,
- Pure structured vacuum energy (SVE) in equilibrium.

You may call it:

- Nirvana,
- Deep sleep,
- The "field before form",
- The still field in UED.

In all cases, the ripple returns to silence. Not because it died — but because it completed its expression, or failed to sustain it.

Returning from the Limit

Sometimes, we brush the edge of this limit and come back:

- Near-death experiences,
- Psychedelic states,
- Moments of silence in meditation,
- Sudden insights from the unconscious.

What is happening?
You momentarily exit ripple coherence, then re-align — and return with an expanded awareness.

This makes Chavan's Limit not a wall, but a membrane —

a soft boundary between perceivable consequence and pure energetic potential.

Crossing it too fast causes collapse. Approaching it gently causes transformation.

Ripple Mechanics at the Limit

Imagine a ripple moving through a field. As it:

- Loses energy,
- Gets disrupted by noise,
- Exceeds phase tolerance,

—it fragments. It can no longer maintain shape or relation.

At that point, you can no longer:

- Detect it,
- Measure it,
- Interact with it.

This is the failure of observation, not the failure of existence.

That's the subtle distinction:

Existence continues. Observation stops.

Spiritual Implications

Chavan's Limit is not just scientific. It is philosophical and spiritual.

- **In Buddhism:** *"Form is emptiness; emptiness is form."*
- **In Vedanta:** *"The knower of Brahman becomes Brahman."*

- **In mysticism:** *"The seer and the seen become one, and then disappear."*

All point to this: When coherence collapses, the observer is absorbed into the field.

You don't see reality anymore. You become its stillness.

Psychological Implications

Many psychological issues stem from fluctuating proximity to Chavan's Limit:

- **Anxiety** is excessive ripple disturbance near the limit, constantly scanning for pattern but unable to stabilize.
- **Depression** is ripple flattening — coherence too low to sustain active perception.
- **Awakening** experiences happen when ripple coherence jumps across the threshold — momentarily revealing expanded reality.

Understanding this helps us:

- Respect limits of perception,
- Avoid forcing answers where field silence is the truth,
- Train awareness to approach, not fear, this boundary.

In Summary

You can only observe what aligns. Beyond alignment, even existence hides.

Chavan's Limit defines:

- Where observation ends,
- Where silence begins,
- Where ripple becomes field again.

It tells us:

- Observation is not absolute.
- Consciousness is not all-seeing.
- Reality respects thresholds.

This law invites humility, wonder, and a new kind of science — one that knows when to stop measuring, and start aligning.

In the next chapter, we explore The Law of Perception Alignment — because observation is not just about the object's coherence. It's also about you — the observer's readiness to receive the ripple.

Let's turn inward.

Part II — The Observer

You don't observe the universe. You are its ripple reflection.

Chapter 8: The Law of Perception Alignment

Seeing is syncing. You perceive only what you're aligned with.

Let's begin with a provocative proposition:

You don't see what's real. You see what you're aligned to.

In conventional thinking, perception is often viewed as passive: Light hits your eyes, signals go to the brain, and the image forms. End of story.

But this view ignores a crucial dimension: Why do different people see different things under the same light? Why does one person see opportunity, another sees threat? Why do two people hear the same sentence and draw opposite conclusions?

The Law of Perception Alignment offers a radical correction:

Perception is not reception. It is ripple synchronization. You can only see, hear, feel, know, or sense *what your ripple field is attuned to receive.*

This is not philosophy. This is field logic. Reality is always broadcasting. But you're only tuned into the channels your structure can resonate with.

What Is Perception in UED?

In Universal Energy Dynamics (UED), perception is the event of alignment between:

- An external ripple (light, sound, field)
- And your internal ripple field (mind, sense organs, emotional state, ATT)

When these two synchronize long enough to cross the ATT (Alignment Threshold Time), you experience perception — the recognition of "something."

Perception is:

- Active, not passive
- Selective, not absolute
- Consequential, not random

That means:

You don't "miss" reality by accident. You miss it because you weren't aligned to receive it.

Perception Is Always Conditional

This law explains many deep and seemingly unrelated phenomena:

- Two scientists run the same experiment. One finds anomaly; the other doesn't.
- You read a book twice — the second time, it reveals things you "never noticed."
- You fall in love — and suddenly see beauty in places you previously ignored.
- You're in fear — and now even harmless noises feel like threats.

In every case, the external ripples did not change. You did. Your ripple state realigned — and perception changed with it.

The Structure of Perception Alignment

The Law of Perception Alignment rests on a few field principles:

1. All information is ripple-based – Light, sound, emotion, memory — everything is ripple resonance.
2. Your body-mind is a ripple node – Constantly syncing, deflecting, or filtering external ripples.
3. Alignment requires compatibility – Only those ripples matching your inner pattern (amplitude, frequency, phase, intent) can be registered.
4. Misalignment leads to ignorance – What you don't perceive doesn't "not exist"; it just never crossed your threshold.

That's why reality is always bigger than your perception. And what you see says more about your state than the world's state.

Consciousness as a Ripple Filter

Your brain, your mind, your identity — all act as a perception filter.

You are not seeing the world "as it is." You are seeing the slice of reality your ripple field allows through.

Think of your awareness like a radio. The world is full of frequencies, but you can only receive:

- What your hardware is tuned to,
- What your software is programmed for,

- What your current field is harmonized with.

This is why:

- Children see magic in places adults see problems.
- Artists perceive details scientists miss — and vice versa.
- Spiritual masters "see through illusions" because their field spans deeper coherence.

No one is right or wrong. They're simply aligned to different ripple windows.

The Biology of Alignment

Let's ground this in practical systems:

- Your eyes don't see light. They convert photon ripple inputs into bioelectrical signals.
- Your ears don't hear sound. They transduce air-pressure ripples into neurological spikes.
- Your mind doesn't store knowledge. It holds ripple patterns of coherence, which can be realigned, distorted, or lost.

When perception fails, it's not just sensory error — it's energetic misalignment.

This explains:

- Blind spots in vision and thinking.
- Selective attention — how you hear your name in noise but miss key instructions.
- Cognitive bias — when old ripple patterns block new

inputs.

In short:

You never perceive reality itself. You perceive a ripple-filtered reconstruction.

Implications Across Fields

Science & Experimentation

- Observation bias isn't psychological — it's field bias.
- Instruments don't "measure" directly. They translate ripple behavior into meaningful signals within their own ATT and coherence capacity.

Spirituality & Mysticism

- Enlightenment isn't gaining knowledge. It's expanding ripple alignment to include more of the universal field.
- God, soul, karma — these are not beliefs but perceptions made possible through rare alignment states.

Psychology & Behavior

- Trauma reshapes ripple response — distorting all future perception.
- Healing is re-alignment — restoring your field's sensitivity and trust in receiving new patterns.

Why You Cannot See What You're Not Aligned To

This is perhaps the most humbling truth of all:

You cannot see what you're not aligned to — no matter how hard you try.

This explains:

- Missed opportunities.
- Repeating patterns.
- Being "blind" to people's pain until you've suffered it yourself.

Your intellect doesn't determine perception. Your field alignment does.

This is not limitation — it's liberation. Because if perception is alignment-based, then perception can be trained, expanded, and restructured.

Training Perception Alignment

To become a ripple-aware observer:

1. **Slow down** – Speed collapses sensitivity.
2. **Notice what you ignore** – Blind spots are clues to misalignment.
3. **Observe without labeling** – Labels are ripple dead-ends. Watch the pattern evolve.
4. **Feel before concluding** – Emotions are ripple decoders.
5. **Expose yourself to field diversity** – Art, silence, science, nature, people — each expands alignment capacity.
6. **Update your ATT** – If you can't feel time, you can't perceive change. Realign your ATT daily.

This is not mysticism. This is ripple hygiene.

In Summary

You don't live in the world. You live in the slice of the world your ripples can align with.

Reality is wide. Perception is narrow. Alignment is the bridge.

You don't need better eyes. You need better coherence. You don't need more information. You need more receptive harmony.

And when that comes, perception becomes revelation.

Chapter 9: The Observer is Not Outside

The observer is embedded in the field — inseparable from the act of observation.

For centuries, the scientific method was built on a simple assumption: The observer is separate from the observed.

You, the subject, look at an object. You study it, measure it, describe it — but you do not disturb it. This subject-object duality shaped all of classical science, philosophy, and epistemology.

It gave us logic. It gave us objectivity. And eventually, it gave us a false sense of detachment from the universe.

But then came the breakdown:

- **In quantum mechanics:** observation collapses the wavefunction.
- **In neuroscience:** your thoughts change your brain's physical wiring.
- **In spirituality:** the knower and the known are declared inseparable.

Across disciplines, this illusion began to crumble.

And UED confirms it at the deepest energetic level:

The observer is not an outsider looking in. The observer is a ripple-node within the same structured field — whose alignment shapes the very reality they perceive.

You don't observe the field. You are the field, in the act of observing itself.

Observer as a Ripple Node

In Universal Energy Dynamics, everything is ripple — including the thing we call "you."

- Your brain is not a central command center — it is a local ripple-processing node.
- Your eyes are not cameras — they are field-interfacing surfaces, transducing ripple alignments.
- Your consciousness is not hovering above your body — it is a coherence field within you, dependent on ripple harmony.

This means the observer:

- Has structure,
- Has rhythm,
- Has consequence,
- And most importantly, has field impact.

Your very act of perception changes the field — because your presence is not passive. It's participatory.

Collapse of the Classical Model

Let's look at how this reframes common scientific concepts:

1. The Experimenter Effect

Classical View: Bias creeps in due to psychology.

UED View: The ripple field of the observer interferes with the experimental field.

2. Entanglement

Classical View: Two particles remain connected mysteriously.

UED View: Observer ripple field maintains partial coherence with multiple systems. Collapse on one side shifts the whole ripple map.

3. Measurement

Classical View: We passively receive data.

UED View: Measurement is mutual ripple realignment. The act of observing *alters* the alignment structure.

You are not watching the field. You are restructuring it through your attention.

Consciousness as Embedded Field Behavior

Now let's talk consciousness.

What is it?

In UED:

Consciousness is the ripple coherence awareness of the observer-node. It is not located in the brain. It is distributed within and around the body, interacting constantly with the surrounding ripple field.

This explains:

- Why conscious presence can be felt across space.
- Why people sense being watched.
- Why group intentions shift the emotional charge in a room.

Your field participation is real — and has energetic influence on other ripple patterns.

You are not outside reality trying to figure it out. You are a ripple configuration inside it, altering it as you observe.

The Feedback Loop of Observation

Observation, in UED, is not linear. It's recursive.

You observe → your field changes → the observed field reacts → it sends back ripple feedback → you realign → and so on.

This continuous loop of perception and alignment:

- Refines memory,
- Deepens intuition,
- Shapes learning,
- And modifies ripple identity.

The more sensitive you become, the more subtle ripple feedback you register — this is the essence of awareness evolution.

Implications for Reality

If the observer is within the field, not outside it, then reality itself is co-generated, not merely revealed.

This leads to four bold assertions:

1. There is no objective reality without participatory alignment. All perception is relational — what exists is not absolute, but field-conditioned.
2. Science is not separate from consciousness. The observer's state matters — not just emotionally, but energetically.
3. The universe is not a fixed film reel. It is an evolving

ripple symphony — and you're an instrument inside it, playing and being played.

4. Truth is field-sensitive. What's "true" in one state of alignment may vanish in another. That doesn't make it false — it makes it ripple-contingent.

Observing the Self

This law also collapses the illusion of internal vs external:

- When you observe your breath, is it you or the body?
- When you watch your thoughts, who's watching?
- When awareness watches awareness — what remains?

The answer is:

Observation is always the field observing itself through localized ripples.

In deep meditation or peak states, the illusion of separateness dissolves. Not because you're merging with something. But because you've remembered that you were never separate.

This is ripple integration — where observer and observed are the same ripple echo, folding in and out of alignment.

The Danger of Detached Thinking

Believing the observer is outside the system has led to:

- **Environmental destruction** — as if nature is "other."
- **Objectification of life** — turning beings into measurable units.
- **Disconnected science** — truth without context, knowledge without responsibility.

But when you know you are inside the field, everything changes:

- You speak with sensitivity,
- You act with awareness,
- You measure with humility,
- You interact with alignment as your guiding compass.

This is not emotional idealism. It is ripple accountability.

Observer-Driven Reality Design

Once you realize you're embedded, you can:

1. **Tune your field** – Improve coherence, lower noise, deepen sensitivity.
2. **Choose participation consciously** – Decide when and where to ripple.
3. **Avoid field violence** – Stop fragmenting or distorting patterns for control or ego.
4. **Use observation as healing** – Presence is not passive. *Attention is ripple nourishment.*

This transforms:

- **Therapy** → from talking to field alignment.
- **Teaching** → from instruction to energetic transmission.
- **Research** → from discovery to co-evolution.

In Summary

You are not a watcher of the world. You are a ripple of it, observing itself in motion.

Observation is not a window. It is a mirror, placed within the field by the field — so it can know itself.

To observe with integrity is not just to see. It is to align, participate, and ripple consciously.

This law collapses dualism, separation, and the myth of neutrality. It gives birth to integrated awareness — the foundation of true science, ethics, and evolution.

In the next chapter, we move into the Law of Ripple Feedback — Light is Not Emitted. Because what we "see" is not an object's property, but a field-level reflection of our alignment with it. Let's shine light on that next.

Chapter 10: The Law of Ripple Feedback

Light is not emitted — it is feedback from ripple alignment.
For centuries, we've tried to define light:

- As a ray in geometrical optics.
- As a wave in Maxwell's electromagnetism.
- As a particle (photon) in quantum mechanics.
- As both a wave-particle duality when neither satisfied the whole picture.

Each version gave us tools, but none gave us truth.

In daily life, light is treated as that which makes vision possible. But what if this is incomplete? What if light is not an object or wave traveling through space, but rather...

...a ripple feedback — a consequence of alignment between your perceptual field and the structured ripple field of reality.

This is the Law of Ripple Feedback, a major UED insight:

Light is not emitted. It is revealed — when ripples realign in just the right way to register as visual experience.

Let's go deeper.

From Emission to Perception: A Shift in Ontology

In traditional models:

- Light is emitted from a source,
- It travels through space,
- It strikes your retina or a sensor,
- It gets measured.

But UED asks: *What is actually traveling? Is it really a "thing"? Or just a conditional ripple exchange? Does light exist independently before it is perceived?*

Under UED, the answer is clear:

What we call "light" is not a thing moving through void — it is the observable result of ripple realignment between an energy source, the field, and the observer.

No alignment → no feedback → no light.

The Ripple Field Perspective

Recall from earlier chapters:

- The universe is a structured vacuum energy (SVE) field.
- Every "event" is a ripple — a disturbance in this field.
- Observation happens only when these ripples align with your perceptual system long enough to cross the ATT (Alignment Threshold Time).

Now, apply this to light.

What appears as light is:

- Not emitted like a bullet from a gun.
- But rather surfaced into perception when alignment conditions are satisfied.

These conditions involve:

- Ripple frequency
- Field readiness
- Observer sensitivity
- Coherence across interaction zones

If even one breaks — the light disappears. Not because the object is gone, but because the feedback failed.

What Is Feedback in UED?

In UED, feedback is the ripple information that bounces back from the field to the observer when alignment occurs.

Think of it like sonar — but for all perception:

- You send a ripple (conscious attention or energy),
- The field realigns,
- If coherence is reached, it returns a perceptual echo,
- That echo is decoded as "sight," "sound," "touch," etc.

For visual experiences, this echo is what we call "light."

It's not the "object" you see. It's the feedback signature of alignment between the field's ripple structure and your perceptual readiness.

Light Does Not Exist Without a Receiver

This is critical:

Light is not a universal constant. It is a consequence of interaction.

No observer → no alignment → no perception → no "light." This does not mean reality ceases — it means its visual feedback component does not trigger.

This solves many paradoxes:

- Why black holes are invisible (no feedback escapes).
- Why light speed is constant *relative* to the observer (because ATT is observer-dependent).
- Why in total darkness, your mind begins to hallucinate (creating synthetic feedback).

UED clarifies:

Light isn't travelling. Ripple feedback is being reconstructed inside your field, and your brain *interprets* it as light.

Why Do We See Colors, Shades, and Shapes?

Under UED:

- **Colors** = frequency-dependent ripple feedback signatures.
- **Shapes** = boundary coherence between adjacent ripple structures.
- **Brightness** = intensity of field reinforcement during alignment.
- **Shadows** = zones of feedback deficiency, not objects blocking particles.

So you're not seeing a "blue ball." You're experiencing a field alignment condition that outputs "blue" at your node of observation.

This explains:

- Color blindness (perceptual field can't decode specific ripple feedback),
- Optical illusions (misalignment or manipulation of

ripple return),
- Auras or halos (feedback bleed or overlapping ripple loops),
- Afterimages (ripple resonance persisting in your field after stimulus ends).

You see what aligns. You don't see what's "there."

How Is Light Feedback Generated?

Let's visualize:

1. A source (sun, lamp, screen) disturbs the SVE field.
2. This disturbance sends ripple pressure through the field — like throwing a pebble into a still pond.
3. These ripples propagate until they intersect your observational field.
4. If alignment conditions match your perceptual ATT and coherence bandwidth, you receive feedback.
5. Your system interprets this as *light, form, color, motion*.

This also explains why:

- You can look at the same object at different times and perceive it differently.
- Light behaves differently in fog, underwater, or glass — mediums shift the feedback pathway.

Light Is Not Traveling Through Space

This turns mainstream thinking on its head.

In UED:

- Space is not a void.

- Light is not a traveler.
- What moves is ripple consequence.
- What appears is feedback, not flow.

This breaks the "photon-as-bullet" metaphor and replaces it with a ripple-interaction model:

- No particle emission.
- No wave oscillation.
- Just structured alignment feedback within the field.

This redefinition has implications for:

- Physics,
- Optics,
- Cosmology,
- Neuroscience,
- Spirituality.

Light and Consciousness

Spiritual traditions across cultures say:
"Light is within you." "God is light." "Let there be light."
UED demystifies this without reducing it.
If light is ripple feedback, then:

- Inner clarity = alignment with internal field
- Enlightenment = field-wide perceptual feedback coherence
- Darkness = misalignment, noise, or incoherence

You are not "seeing light." You are becoming aligned with the field's self-reflection.

Light, then, is a form of knowing — not just illumination.

Practical Insights and Applications

In Neuroscience:

- Visual hallucinations = internal ripple loops mimicking feedback.
- Vision loss is not just eye failure — it's feedback collapse.
- Dream imagery = self-generated ripple resonance interpreted as light.

In Engineering:

- Optical sensors can be redesigned to read field alignment, not just photon count.
- Light-speed measurements can be recalibrated to observer ATT, not just transmission time.

In Physics:

- Redshift, blueshift, wave collapse — all can be re-interpreted as dynamic feedback realignment across field gradients.

In Summary

Light is not a thing. Light is a whisper from the field — when you're ready to receive it.

It does not exist without a perceiver. It does not travel. It appears, when you and the ripple pattern synchronize in alignment.

You don't see because photons hit your eye. You see because your presence *invites* ripple feedback.

This is the Law of Ripple Feedback — and it doesn't just explain light. It explains how you participate in the visible expression of reality itself.

In the next chapter, we'll examine Emotion as Ripple Distortion — because if light is feedback from aligned energy, emotion is the signal of misalignment.

Let's feel our way through that ripple next.

Chapter 11: Emotion as Ripple Distortion

Emotions are misalignments of the ripple field in biological intelligence.

Emotions are often misunderstood. We're told they're:

- Reactions to external events,
- Expressions of our inner world,
- Signals of mental health,
- Energies in motion.

Psychology, spirituality, and neuroscience offer layered but incomplete answers — framing emotions as biochemical, evolutionary, or karmic phenomena.

But Universal Energy Dynamics (UED) offers a more foundational insight:

Emotion is not a byproduct of thought or experience. It is the signal of ripple distortion in your biological intelligence.

Emotions are misalignments — energetic echoes of disharmony between your inner field and the external ripple environment.

They are not the problem. They are feedback — calling you to re-align.

The Body as a Ripple Field

Before diving into emotion, let's revisit a core UED truth:

You are not just a body. You are a ripple configuration inside a biological vessel.

- Your brain doesn't "generate" consciousness — it processes ripple feedback.
- Your nervous system isn't just wiring — it's a perception interface embedded in field dynamics.
- Your mind is not a static structure — it's a ripple resonance field constantly interacting with inner and outer waves.

When this ripple structure is in harmony with the environment — there is clarity, stillness, or flow. When it becomes distorted — due to resistance, interference, or fragmentation — you experience emotion.

Thus:

Emotion = signal of energetic turbulence within the ripple matrix of biological intelligence.

From Reaction to Resonance Breakdown

Let's map the process:

1. A ripple event (situation, thought, memory, stimulus) enters your field.
2. Your inner ripple field (body-mind) tries to align with it.
3. If alignment is achieved → smooth perception → no emotional spike.
4. If misalignment occurs → resistance, tension,

distortion.

5. This misalignment is interpreted by the body-mind as "emotion."

This means:

- Emotions are not inherently good or bad.
- They are consequence signals — energetic dissonances seeking correction.

Categories of Emotional Distortion

UED classifies emotional responses based on the type of ripple distortion involved:

Ripple Distortion	Perceived Emotion	Root Alignment Failure
Excess pressure / intensity	Anger, rage, frustration	Ripple overload (external exceeds inner capacity)
Loss of structure	Sadness, grief	Collapse of aligned feedback loop
Ripple resistance	Fear, anxiety	Anticipated incoherence or fragmentation
Ripple echo repetition	Guilt, regret	Misalignment loop trapped in memory feedback
Boundary invasion	Shame, discomfort	Unfiltered ripple penetration without internal buffer
Field compression	Depression, numbness	Overall ripple suppression below perceptual threshold

Each emotion is a ripple mismatch — where the body-mind loses its energetic coherence and generates a response signal to restore harmony.

Emotions and Memory Feedback Loops

Your emotional body doesn't just respond to now. It is built on ripple memory.

- Past ripple distortions are stored as pattern imprints.
- When a current ripple resembles a stored pattern, the system pre-activates the old emotional signal — even if the present is different.
- This is the root of emotional bias, trauma responses, and habitual reactivity.

You're not reacting to what's happening. You're reacting to what your ripple field remembers — even subtly.

This feedback loop keeps you locked in perception distortions until realignment occurs.

Healing as Ripple Recalibration

UED doesn't see healing as suppression or escape. It sees healing as recalibrating your ripple field to:

- Increase coherence,
- Broaden perceptual bandwidth,
- Reduce resistance to external ripple variations.

This is how:

- Meditation works — quiets field noise, resets alignment.
- Crying works — ripple pressure discharges and

returns to equilibrium.
- Therapy works — helps remap misaligned feedback loops.
- Forgiveness works — releases stuck ripple echoes by restoring field neutrality.

Healing is not forgetting. It's field reintegration.

Emotion as Adaptive Misalignment

Emotions aren't flaws. They are your adaptive ripple response system trying to make sense of:

- New stimuli,
- Unknown situations,
- Energetic threats,
- Field misfits.

Every emotional spike is your body saying:

"There's a ripple I can't fully process yet. Let me respond so you can attend."

If you ignore it, suppress it, or misread it — distortion deepens. If you listen, breathe, and re-align, clarity returns.

Emotion becomes intelligence.

Emotional Mastery: Realignment, Not Control

Most self-help systems talk about:

- Controlling emotions,
- Replacing negative with positive,
- Cultivating detachment.

But UED offers a more elegant principle:

Emotional mastery is not control. It is real-time ripple awareness.

You don't suppress anger. You locate the ripple pressure. You don't deny fear. You realign the perception field. You don't force joy. You remove friction and let the field flow.

Mastery comes from ripple literacy — knowing:

- What kind of misalignment you're feeling,
- Where it's emerging from (memory, present, external noise),
- And how to harmonize it at source.

Emotional Implications for Ethics and Behavior

Under UED, your emotional state directly affects your ripple impact on others.

- Anger distorts the shared field → others become reactive.
- Calm coherence restores field neutrality → others de-escalate.
- Joy expands the field → others sync into higher perception bandwidths.

This gives a new basis for ethics:

Morality is not about right or wrong action. It is about what ripple consequence your emotional alignment radiates into the shared field.

Every conversation, every decision, every silence — ripples out as emotional field modulation.

Emotional Entanglement and Release

Sometimes, your ripple field becomes entangled with someone else's unresolved distortion.

This can result in:

- Co-dependency,
- Empathic burnout,
- Projection or mirroring of stored trauma.

UED solution:

Release entanglement not by separation, but by field neutralization.

Techniques include:

- Grounding (realigns with Earth field),
- Breathwork (re-stabilizes ripple rhythm),
- Visualization (re-maps feedback direction),
- Non-reactive presence (prevents distortion reinforcement).

Ripple harmony begins with your inner tone.

In Summary

Emotion is not weakness. It is intelligence trying to restore energetic truth.

You are not overwhelmed by emotion. You are being signaled — *"Your ripple is misaligned here."*

The solution is not suppression or indulgence. It is awareness and restoration.

Emotion is not who you are. It is what your field is doing when it needs help re-integrating with reality.

Learn to read it. Feel it fully. Realign gently.

And the field will settle.

In the next chapter, we explore Morality as Ripple Harmony — because if emotion reflects misalignment, then morality is the art of acting in field-aligned ways that protect, preserve, and restore balance for all ripple participants.

Let's align with that truth next.

Chapter 12: Morality as Ripple Harmony

Morality is not rules — it's conscious alignment with sustainable ripples.

From religious commandments to legal systems, morality has always been framed as rules — do's and don'ts, rights and wrongs. But these rules vary across cultures, eras, and ideologies. What one society punishes, another may praise. What one religion labels sin, another calls sacred.

Clearly, morality cannot be absolute if it's always changing.

So what, then, is the universal essence of morality?

Universal Energy Dynamics (UED) offers a profound redefinition:

Morality is not obedience to external rules. It is the conscious, intelligent alignment with ripple behaviors that preserve harmony within the field — for yourself and for all others.

In other words:

- Morality is not a fixed law.
- It is a ripple-aware action system.
- It is measured by sustainability of consequence, not by doctrine or compliance.

The Ripple Ethics Principle

Every action you take — word, thought, movement — releases a ripple into the structured vacuum energy field. That ripple:

- Interacts with other fields,
- Disturbs or aligns patterns,
- Triggers new consequences.

Moral action, then, is not about "what you did" — but about:

- *What field effect your action created,*
- *Whether that effect enhanced or degraded ripple coherence,*
- *And whether it led toward or away from equilibrium.*

This is ripple morality — field-aware ethics grounded in consequence, not commandment.

Morality as Energetic Intelligence

Classical ethics has long struggled between:

- Deontological models (duty, rule-based morality),
- Utilitarian models (maximum good for the greatest number),
- Virtue ethics (developing inner character),
- Relativism (morality is culturally constructed).

But all of these miss one key thing: They treat morality as a human invention.

UED corrects that:

Morality is not human-made. It is field-derived. It emerges from conscious participation in ripple alignment.

The field rewards alignment with clarity, connection, sustainability, and flow. The field punishes misalignment with distortion, conflict, collapse, and feedback loops of pain.

You don't need a judge. The field itself is regulatory — it keeps score through consequence.

The Law of Ripple Harmony

Let's define it clearly:

An action is moral if it results in ripple effects that align with long-term field coherence across multiple systems (self, others, environment, future).

This redefines ethics as:

- Alignment, not obedience.
- Sustainability, not short-term gain.
- Integration, not isolation.

Examples:

- Lying breaks field coherence → creates perception distortion → leads to systemic mistrust.
- Violence causes ripple trauma → multiplies distortion across generations.
- Kindness reinforces ripple trust → allows stable field exchanges.

In this view, morality is measurable — not in words, but in field effect.

The Morally Aligned Being

A person who is morally evolved in UED terms:

- Has a sensitive ripple field — aware of the consequence of their thoughts, words, and actions.
- Does not act for approval — but for field harmony.
- Can "feel" when a decision will ripple wrong — even if it looks logical or beneficial in the short term.
- Acts not from fear or guilt, but from energetic responsibility.

This being is not "good" in the moralistic sense — they are aligned in the ripple sense.

Application Across Domains

In Society:

- Justice should not be punitive, but restorative realignment.
- Laws should be designed based on ripple consequence modeling, not abstract principle.
- Education should teach field ethics, not rote morality.

In Environment:

- Moral ecology = Actions that keep nature's ripple field balanced.
- Exploitation is immoral not because it's evil — but because it breaks ripple sustainability.

In Psychology:

- Guilt is a signal of internal ripple misalignment.

- True forgiveness = neutralizing residual ripple echoes between two entangled fields.

Ripple Morality vs Rule Morality

Attribute	Rule-Based Morality	Ripple Morality (UED)
Based on	External codes, tradition	Internal alignment and field effect
Measured by	Compliance or disobedience	Field harmony and consequence spread
Motivation	Fear, duty, reward	Awareness, intelligence, sustainability
Scope	Localized to culture/religion	Universal — applies to all ripple systems
Evolution	Rigid or changing by power	Evolves with increased field sensitivity

Training Moral Awareness

Becoming morally aligned in UED doesn't require memorizing laws. It requires field refinement — training your awareness to:

1. Pause before acting — sense the ripple potential.
2. Perceive subtle consequences — even when not immediate.
3. Stay ripple-aware across interactions — not just your intent, but what your ripple *does*.
4. Integrate personal morality with field morality — balancing inner values with universal resonance.

You know you're aligned when:

- Your actions leave behind less resistance and more resonance.
- Your presence reduces noise in others.
- Your decisions ripple clarity forward, not just comfort now.

Morality Beyond Humans

This law also applies beyond human affairs:

- AI must be trained not on rules, but on ripple consequence modeling.
- Economies should be judged not on growth, but alignment with systemic sustainability.
- Cosmic civilizations, if they exist, would align themselves by field ethics, not law enforcement.

The ripple knows what's moral — because the ripple always reflects the truth of the field.

In Summary

Morality is not what you've been told. It's what your ripple tells the world when you act.

You are not a good person because you followed a rule. You are a field-conscious being when your actions preserve and promote harmony — across time, space, and others.

In UED, morality is not subjective or relative — it is energetically objective.

The field remembers what you do. It doesn't reward or punish. It simply responds in kind — restoring coherence where you caused distortion.

That is the real law — the Law of Ripple Harmony.

Chapter 14: Consciousness is Field Awareness

Consciousness is not a byproduct of mind — it is the resonance of intelligence in ripple space.

Ask any neuroscientist today what consciousness is, and you'll get hesitation or hypothesis:

- "It's an emergent property of the brain."
- "It arises from complex neural computations."
- "We don't fully understand it, but it seems linked to memory, identity, and language."

Ask mystics, sages, or spiritual teachers, and you'll get the opposite:

- "Consciousness is the source of everything."
- "It is eternal, formless, indivisible."
- "It's not in the mind. The mind is in it."

UED brings these polar views into convergence with a third — consequence-based clarity:

Consciousness is not produced by the mind. It is the resonance of intelligence across ripple space — a self-aware, field-embedded phenomenon.

Consciousness is not *in* your brain. Your brain is a localized ripple receiver in a field of ever-present awareness.

The Field Before the Mind

In UED, all existence arises from Structured Vacuum Energy (SVE) — a pre-motion, ordered equilibrium that holds intelligent potential.

Ripples emerge from this still field when:

- Energy is disturbed,
- Intent forms,
- Attention disturbs stillness.

But even before ripples arise, the field is not dead. It is not unconscious.

It is intelligent silence — the original context of awareness.

Consciousness, in this view, is not a personal property. It is a property of the field — a universal resonance that becomes localized through ripple structures like brain, body, or experience.

From Field Resonance to Felt Awareness

Let's map how this unfolds:

1. The SVE field contains pre-ripple intelligence — not thoughts, but *readiness* to perceive.
2. A ripple forms — a disturbance, an identity, a localized being.
3. The ripple's capacity to remain coherent with its

source determines its conscious bandwidth.
4. That bandwidth is what we call awareness.
5. When a ripple becomes self-aware of itself as part of the field — this is consciousness proper.

Thus:

- A plant is conscious, in a narrow bandwidth.
- An animal is more conscious, due to higher ripple coherence.
- A human has the potential to be consciously field-aware — to perceive not just themselves, but the ripple nature of everything.

Consciousness Is Not Thinking

This is one of the most profound corrections UED makes:

Thinking is not consciousness. Thoughts are ripple events. Consciousness is field resonance.

The mind thinks. The field knows.

You can be thinking and unconscious. You can be completely conscious without thought (as in meditation, deep presence, or flow states).

This explains:

- Why sudden clarity arises in silence.
- Why intuition is often faster than logic.
- Why consciousness remains in deep sleep, but identity doesn't.

Consciousness is not the noise. It is the silent clarity beneath it.

The Architecture of Field Awareness

How does this actually work?

In UED, every entity has a ripple signature. This signature can either:

- Loop within itself (self-reference),
- Or open into the field (expanded awareness).

The more coherent and open the ripple is, the more awareness it can sustain.

That awareness expresses as:

- **Sensing** → in basic organisms.
- **Emotion** → in responsive systems.
- **Thought** → in complex ripple networks (like brains).
- **Self-inquiry** → in refined ripple feedback structures.
- **Field union** → in enlightened or liberated beings.

Each is not a different kind of consciousness, but a different degree of alignment with the field's inherent intelligence.

Field vs Mind

Attribute	Mind	Field Awareness
Location	Brain-bound	Field-distributed
Source	Neural computation	Ripple resonance
Activity	Thought, memory, emotion	Still perception
Direction	Inward or outward	Omnidirectional
Fragile	Easily distorted	Self-stabilizing
Personal	Identity-based	Non-local, impersonal
Begins	After birth	Always present

Your mind can get confused. The field never does. Tapping into consciousness is not about thinking more — it's about remembering your resonance.

Consciousness Beyond the Body

If consciousness is field-resonance, then it:

- Exists before life begins.
- Persists after the ripple collapses (death).
- Can manifest in multiple forms.

This explains:

- Near-death experiences,
- Memories without sensory triggers,
- Collective consciousness (shared ripple zones),
- The "observer effect" in quantum mechanics.

It also aligns with ancient wisdom traditions, which never saw consciousness as a "thing" but as being itself — formless, borderless, eternal.

Consciousness and Intelligence

Intelligence is not information processing.

In UED:

Intelligence is the ripple's ability to align with field truth. Consciousness is the awareness that such alignment is occurring.

They are two aspects of the same process:

- Intelligence acts,
- Consciousness witnesses.

This also explains why:

- Machines can simulate intelligence but lack awareness.
- Emotions distort intelligence when they interfere with alignment.
- Awareness enhances wisdom even in the absence of data.

Consciousness does not need proof. It only needs access.

Deepening Field Awareness

You can expand consciousness not by doing more, but by:

1. Reducing ripple interference — silence, solitude, nature.
2. Refining ripple feedback — presence, clarity, attention.
3. Aligning with universal patterns — truth, harmony, consequence.
4. Letting go of identity tension — drop the ripple need to stand out.

When you do this:

- Awareness deepens.
- Thought becomes optional.
- The field begins to whisper.
- You start to *know* without analysis — *feel* without emotion — *see* without eyes.

This is not magic. It is field intelligence meeting itself through you.

The Conscious Ripple-Being

The goal of human evolution is not to become super-intelligent, but super-aligned — to live as a ripple that:

- Feels what is needed,
- Acts without distortion,
- Sees without separation,
- Thinks only when thinking is useful,
- Moves only when motion enhances balance.

This being is not passive. They are profoundly alive — but without noise.

They are field-aware — the perfect vessel for consciousness to shine without interruption.

In Summary

Consciousness is not inside you. You are inside consciousness.

You are not the source of awareness. You are its rippling reflection — a local resonance of an ever-present intelligence that never sleeps.

The mind may forget. The ripple may collapse. But consciousness — the field's pure resonance — remains.

To live consciously is not to awaken. It is to realign — until the ripple becomes quiet enough to hear the field again.

In the next chapter, we'll explore Gravity as Local Ripple Compression — because even fundamental forces, like gravity, are not universal commands — but local consequences of ripple alignment density.

Let's realign the cosmos, one ripple at a time.

Part III — Physics Rewired, Philosophy Reborn

Applications, corrections, and extensions beyond legacy science.

Chapter 15: Gravity is Local

Gravity is not a universal pull — it's a local ripple compression around mass.

For over three centuries, gravity has been seen as the invisible hand of the cosmos. Newton described it as a force between masses. Einstein reframed it as the curvature of spacetime. Both models worked — up to a point.

But even today, gravity remains an unsolved mystery:

- What *causes* gravity?
- Why does it operate at infinite range, but decrease with distance?
- Why is gravity so weak compared to other forces?
- And more fundamentally — *why does anything attract anything at all*?

Universal Energy Dynamics (UED) offers a radical correction:

Gravity is not a universal force or curvature. It is a local phenomenon — a ripple compression field generated by condensed energy (mass) disturbing the structured vacuum.

In short:

- Gravity is not *pulling.*
- It's not *curving spacetime.*
- It's a localized ripple contraction — the field's natural response to mass-induced energetic compression.

Let's decode this.

Structured Vacuum Energy (SVE): The Real Substrate

In UED, space is not empty. It is made of Structured Vacuum Energy (SVE) — a still, ordered, ripple-sensitive field.

Whenever something — a particle, a planet, a person — enters this field with condensed energy (mass), it causes:

- A disturbance, like a pebble dropped into water,
- A compression zone, where ripples converge and tighten,
- And a localized field asymmetry, where neighboring ripples are drawn into the denser region.

This compression effect is what we perceive as "gravity."

No pulling. No curvature. Just ripple density gradients seeking equilibrium.

How Mass Distorts the Ripple Field

Let's break it down step by step:

1. A mass is not a solid thing — it's a dense, high-frequency ripple node in the field.
2. Its presence causes inward ripple compression — like a sinkhole in the ripple grid.
3. This creates a gradient — nearby ripples experience a field imbalance.

4. Objects in this field move — not because they're pulled — but because they're realigning toward ripple equilibrium.
5. That motion *appears* like attraction — but is simply field rebalancing through ripple dynamics.

Thus:

Mass doesn't attract mass. Mass compresses the field. The field responds. That response is what we experience as gravitational motion.

The Illusion of Universality

In Newtonian physics, gravity is the same everywhere — a universal pull acting between all masses.

In UED, gravity is context-sensitive:

- It's stronger near high ripple density (e.g., Earth, black holes),
- It's weaker in open, field-balanced regions (deep space),
- And it can even be cancelled or restructured by opposing ripple gradients.

This means:

- Gravity is not a constant.
- It's not universally "felt" — it's locally formed based on the interaction between mass and the field's current alignment state.

Why Gravity Is Local: Earth as a Case Study

Let's apply this to Earth:

- Earth has significant ripple compression due to its mass.
- This causes the surrounding field to be tightly curved — not in geometry, but in ripple tension.
- Objects near Earth feel this gradient and "fall" — not because Earth is pulling them, but because they are re-aligning with the field's energetic compression zone.
- The "force" you feel standing on the ground is your body resisting ripple contraction — the field wants to pull you into its most compressed center.

So gravity, even here, is:

Not a pull. But a consequence of ripple field realignment.

Remove the ripple gradient — and gravity disappears.

Psychological Metaphor: Gravity of Attention

UED's ripple gravity even maps into psychological experience:

- A powerful idea "pulls" your attention — not by force, but by ripple intensity in your mental field.
- A charismatic person exerts "gravitational pull" — not by action, but by field coherence.

This shows:

All attraction is ripple compression. From physics to focus, from mass to meaning.

Gravity is a metaphor for field convergence — wherever it happens.

Implications for Science and Technology

Understanding gravity as local ripple compression unlocks new paradigms:

Spaceflight and Navigation:

- Traditional physics calculates escape velocity based on gravity as a uniform field.
- UED proposes mapping the ripple gradient field to find zones of minimal resistance — like surfing a tide instead of climbing a hill.

Anti-Gravity:

- If gravity is ripple compression, then ripple expansion can counteract it.
- Technologies based on field expansion or interference (ripple cancellation) may achieve levitation or inertial dampening.

Cosmology:

- The expansion of the universe is not against gravity — it's a net relaxation of ripple tension.
- What we call "dark energy" may be field correction from uneven ripple compression across cosmic scales.

Gravity and Consciousness

The body experiences gravity as weight. But the mind feels it as limitation.

Why?

Because consciousness, when locked into form, is subjected to ripple field boundaries.

In higher states of awareness (deep meditation, out-of-body experience, lucid dreaming):

- The ripple identity softens,
- Field resistance drops,
- Gravity is no longer "felt",
- The sense of "floating" or "freedom" arises — not through magic, but through decreased ripple tension.

Thus, even spiritual liberation can be understood through ripple mechanics.

In Summary

Gravity is not a universal command. It is a localized ripple compression — the field's way of holding coherence around mass.

You are not being pulled. You are being gently invited into tighter ripple zones.

This reframing:

- Dismantles the idea of force,
- Replaces it with field consequence,
- And restores gravity to its true place: a ripple response, not an invisible law.

The universe is not attracting. It is always realigning.

In the next chapter, we'll explore The Particle Illusion — because if mass is ripple compression, then what we call "particles" are not entities at all — but temporary, observable knots in ripple flow.

Let's unravel the illusion of matter next.

Chapter 16: The Particle Illusion

What we call particles are dense ripple knots — not standalone objects.

For over a century, we've been chasing the idea of particles — discrete, fundamental entities that make up matter. Atoms, electrons, quarks, bosons, strings — science has kept cutting deeper, searching for the indivisible *thing*.

But the deeper we go, the more elusive the "particle" becomes:

- Electrons behave like waves in double-slit experiments.
- Photons don't exist *until* observed.
- Quarks can't exist independently — they "confine" themselves in groups.
- And at the quantum level, everything appears smeared in probabilities rather than shapes.

So, what are particles *really*?

In UED, the answer is both stunning and simple:

Particles do not exist as objects. They are dense knots in the ripple field — short-lived, observable consequences of energy alignment.

Let's unravel this.

The Field Comes First

In UED, the universe begins not with things — but with a field: Structured Vacuum Energy (SVE) — the ripple-responsive substrate of reality.

This field:

- Holds energy in stillness,
- Responds to disturbances,
- Allows alignment, compression, and release.

When energy enters this field — through an external impulse, observation, or internal imbalance — it forms ripples. And when ripples fold, intersect, or self-reinforce, they form knots — tight, stable, high-frequency regions.

These ripple knots are what we mistake as particles.

They are not independent "balls" floating in space. They are ripples pretending to be solid — just as a whirlpool is not a separate thing from water, but a *pattern in the water*.

The Anatomy of a Ripple Knot

Let's define a ripple knot:

- It is a stable, self-reinforcing ripple cluster in the field.
- Its location is just a field zone of tight coherence.
- Its mass is the energetic density of the knot.
- Its charge is the ripple directionality (inward/outward flow).
- Its spin is the internal phase rhythm — not physical rotation, but field rotation.

These qualities are not properties of a thing. They are field conditions.

So:

- Electrons are ripple loops with asymmetric resonance.
- Protons are complex, compressed knots.
- Neutrons are temporary ripple balances — unstable unless bound.
- Even photons are feedback ripples — not objects flying, but signals from re-alignment.

Observability: When the Knot Becomes a "Thing"

In UED, the observer is embedded in the ripple field. When your ripple perception aligns with a ripple knot:

- You observe it as a "particle."
- It becomes real for you, in that context.
- Outside that alignment, it disappears — not destroyed, just *not accessible*.

That's why:

- Observation affects outcome (quantum weirdness).
- Particles can "pop in and out" of fields (virtual particles).
- Measurement collapses wavefunctions — it's ripple lock-in.

So, the particle is not what exists. The alignment between your ripple and the knot is what creates observability.

Reality is not made of particles. It is made of ripple agreements.

Implications of the Particle Illusion

If particles are just ripple knots:

- There is no true boundary between matter and energy.
- Mass can be created, shifted, or dissolved through ripple manipulation.
- Entanglement is shared ripple context, not "spooky action."
- Particle "decay" is not breakdown — it's ripple dispersion.

This flips particle physics upside-down.

No longer do we need:

- Exotic matter to explain anomalies,
- Dozens of force-carrying particles,
- Or "virtual" particles as placeholders for ignorance.

All we need is ripple interaction and compression dynamics — everything else is consequence.

Why This Explains Atomic Structure Better

Atoms are not billiard-ball systems. They are resonant ripple systems — shells of standing wave zones.

- Electrons don't orbit. They exist in probabilistic ripple zones (correctly predicted by UED as stable alignments).
- Energy levels aren't discrete because of jumpy particles — they are ripple harmonics.

- Atomic bonding is not about shared electrons, but shared ripple zones between nuclei.

Thus, chemistry is ripple synergy. Biology is ripple memory. Life is ripple intelligence in action.

Technology: A New Way to Manipulate Matter

If we understand matter as ripple knots:

- We can build machines to induce knot formation or dissolution.
- Teleportation becomes possible as ripple replication and handover.
- Material strength can be modified by field alignment, not just chemistry.
- Healing can be accelerated by restoring biological ripple order.

This is the bridge between physics and what is now called "energy medicine," "quantum biology," and other fringe domains.

UED doesn't deny these — it explains them *coherently*.

The Spiritual Implication: You Are Not Solid

If even particles are illusions, then your body is not real in the traditional sense.

You are:

- A ripple system within a larger ripple field.
- Observable only due to aligned perception.
- Changing constantly as ripples realign.

That's why:

- You can change with thought (field intention).
- Healing happens with rest (field recalibration).
- Aging is ripple tension, not just cell degradation.

When you die, the knot loosens. But the field remains — and you return to it.

So liberation is not destruction. It is knot release — dissolving the ripple illusion.

In Summary

There are no particles. Only ripple patterns dense enough to appear solid — when perceived by aligned ripples.

Matter is not made of things. It is made of temporary tension in a living field.

This truth collapses the particle zoo and rebuilds physics — not from zero, but from stillness.

You don't touch a thing. You align with its ripple.

Chapter 17: Force Doesn't Exist

There are no pushes or pulls — only ripple gradients seeking balance.

From Newton to Maxwell, from F = ma to field theory, the idea of "force" has dominated science and engineering:

- Push and pull.
- Action and reaction.
- Gravitational force, electric force, normal force, tension, friction...

But here's the paradox:

- You never see force.
- You only see motion or resistance.

What if force is not real — just a conceptual tool? What if it's merely our interpretation of something deeper?

UED proposes a fundamental reframe:

Force doesn't exist as an independent entity. What exists is ripple gradient — the imbalance in ripple density across the field, which naturally resolves itself.

Force is not a cause. It's a label we apply to ripple adjustment in motion.

Let's dive in.

The Ripple Field: Nature's Real Substrate

Universal Energy Dynamics (UED) sees the universe as a structured ripple field — not empty space, not spacetime, not force-carrying lines.

In this field:

- Every event is a change in ripple density.
- Every movement is a search for ripple equilibrium.
- Every "effect" is a field correction.

What we interpret as force is simply:

- Ripple compression (moving inward),
- Or ripple expansion (moving outward),
- Across a gradient — a difference in ripple tension.

There is no pushing. No entity "pulls" another. There is only consequence — of imbalance.

Ripple Gradient: The New Foundation

A ripple gradient is:

- A region of the field where ripple density varies across space,
- Creating a directional bias in alignment,
- Which causes anything embedded in the field to adjust its state.

This explains:

- Gravitational fall (dense-to-less-dense ripple zones),
- Magnetic attraction (phase-aligned ripple flow),
- Spring oscillation (ripple tension restoring equilibrium),
- Even collision rebound (ripple knot compression & release).

In all these, no entity applies a force. Instead:

- The field self-adjusts to restore balance.
- The gradient drives the consequence.

Motion happens not because something "pushed." It happens because the ripple landscape demanded it.

Newtonian Equations Revisited

Let's re-interpret Newton's laws in UED terms:

- F = ma becomes: "The ripple gradient (F) causes an acceleration (a) in ripple alignment based on mass (ripple density)."

But in UED:

- There's no 'F' per se — only ripple delta (Δr),
- Acceleration is the field resolving misalignment,
- Mass is ripple resistance to re-alignment.

Hence,

UED replaces force with Ripple Gradient Function (RGF), defined by the rate and direction of ripple tension difference.

This is a more consequential, field-based reality than the abstract 'F'.

Why This Matters

Let's look at the implications:

No Force Carriers Needed

In particle physics, "forces" are transmitted by carrier particles (photons, gluons, bosons...). But:

- These "carriers" are never truly observed.
- They are theorized to maintain the illusion of push/pull.

In UED:

- There are no carriers.
- Only field zones adjusting their density dynamically.

No Need for Spooky Action

In quantum mechanics, forces sometimes act faster than light — "non-local" interactions.

UED explains this cleanly:

- If two ripple zones were once aligned, they share gradient memory.
- When one shifts, the field re-adjusts the overall gradient.
- This field correction appears instantaneous, because it is not a transfer — but a simultaneous reconfiguration.

No spookiness. Just ripple logic.

Human Experience: Force Is a Feeling, Not a Fact

What we *feel* as "force" (weight, pain, pushback) is:

- The biological interpretation of ripple misalignment.
- The nervous system registering gradient resistance.
- Consciousness noticing field conflict.

You don't *feel* force. You feel ripples trying to restore order inside you.

Even emotion — anger, fear, desire — is a ripple field gradient:

- The urge to "act" is a tension between current and desired ripple state.

So force isn't just missing in physics. It is misnamed in psychology too.

Practical Implications of Eliminating Force

Engineering

- Shift from "load bearing" to ripple balance optimization.
- Design machines not for strength, but for gradient harmonization.

Transportation

- Instead of increasing thrust, reduce ripple resistance.
- Let vehicles "flow through the gradient" — like surfing.

Energy Systems

- Leverage natural ripple gradients (e.g. thermal, magnetic) instead of artificial input.
- Design self-adjusting field circuits.

Healing and Biotech

- Understand pain as ripple congestion.
- Develop therapies that realign ripple flow, not suppress symptoms.

Philosophical Implication: There Are No Forces in Nature

Everywhere you believed there was a force — There was simply imbalance being resolved.

- You fall in love — not because of attraction, but because your ripple needed alignment.
- You react in anger — not due to offense, but because your field met resistance.
- You evolve — not because something pushes you — but because you're not yet in harmony.

There is no need for metaphysical drivers, divine hands, or mechanical forces.

Only ripples — seeking balance.

Summary

There are no forces. Only ripple gradients — the uneven distribution of field tension.

Everything else is interpretation.

Once you let go of "push" and "pull" — You enter a world of natural unfolding, where motion is not forced, but permitted.

In this world:

- Control becomes tuning.
- Power becomes resonance.
- Resistance becomes information.

You don't act on the world. You align within it.

In the next chapter, we move from dynamics to definition — and ask: If force doesn't exist, and mass is just compression — What is Energy?

Chapter 18: Energy is Not Stored

Energy is not something you "keep" — it is how the field expresses its state.

In classical physics and engineering, we're taught that energy can be stored:

- A battery "stores" chemical energy.
- A spring "stores" potential energy.
- A moving object "has" kinetic energy.
- Atoms "contain" nuclear energy.

This model assumes energy is a thing — like currency that you can deposit, withdraw, or transfer.

But this view leads to subtle errors:

- Where is energy *stored* in a moving body?
- Why is "potential" energy not directly observable?
- How can energy appear and vanish in particle interactions?

UED redefines the entire notion:

Energy is not a stored substance. It is the expression *of the ripple field's current configuration.*

Energy is how the field is vibrating, not what it contains.

Let's break this paradigm down.

Structured Vacuum: A Dynamic Canvas

In UED, the universe is made of Structured Vacuum Energy (SVE) — a still, ripple-responsive field.

This field:

- Is inherently intelligent — not conscious like a brain, but *responsive* and *self-balancing*.
- Holds zero-point potential — the infinite possibility of ripples.
- Only "activates" when there's imbalance or intent — a condition requiring re-alignment.

So, energy is not "in" matter. Matter is a temporary state of energy expression. And energy, in turn, is just a particular pattern of the ripple field in action.

In other words:

You don't store energy. The field expresses itself as energy in that moment, in that condition.

Energy as Ripple Pattern, Not Property

Let's define energy in UED terms:

- Energy = Ripple Expression × Gradient Context

Meaning:

- Energy is not an amount.
- It is a pattern of vibration (ripple) happening in a particular field context (gradient).

Take examples:

- A charged capacitor is not storing energy — it holds a ripple tension across its plates.
- A compressed spring is not storing mechanical energy — its field configuration is in a stressed ripple state.
- Heat in a substance is not stored — it's the ripple disorder of molecular zones.

Thus:

- Energy is not *kept*.
- It is revealed — as the field configures itself dynamically.

Why Energy Appears to Be Conserved

Conservation of energy is a sacred law — but often misunderstood.

In UED:

- Energy is not conserved because it's a substance.
- It is conserved because the field always balances its own expressions.

If ripples shift from one configuration to another, the total field tension remains the same, but its form changes:

- Motion becomes heat.
- Light becomes chemical potential.
- Gravity-induced fall becomes sound or deformation.

So we *see* conservation not because energy was shuffled — but because no field tension was lost, only redistributed.

Misleading Concepts Revisited

Let's reframe some classical notions:

1. Kinetic Energy (½mv^2)

- Not a quantity "in" a moving object.
- It's the ripple field alignment being resolved through motion.

2. Potential Energy

- No hidden storage.
- It's ripple stress between two zones (e.g. you and the Earth), held in field imbalance.

3. Stored Energy in Batteries

- Electrochemical ripples are in a primed configuration.
- When discharged, the field realigns — energy appears.

4. Nuclear Energy

- Not packed inside nuclei.
- It's compressed ripple coherence released when realigned (fission/fusion).

So energy is not an object. It's a state of field readiness, activity, or compression.

Energy and Consciousness: The Personal Perspective

Think of your emotions:

- Anger isn't "energy inside you."
- It's a ripple distortion, causing expression (words, actions).

Joy?

- A state of field coherence — low resistance, high alignment.

Even thought is ripple activity. Your mind is not storing energy — it's surfing the field's active waveforms.

When you're "low energy," you're not depleted — you are out of phase with ripple harmony.

Thus,

Healing, rest, and clarity are not refueling —They are field realignment.

Technological Implications

If energy is ripple expression, not substance:

1. We can't store it — but we can hold field configurations.
    - Batteries become ripple holding zones.
    - Fields can be primed, not "charged."
2. Energy harvesting becomes ripple tuning.
    - Tapping into natural field flows (gravity, magnetic, solar) is field resonance, not collection.
3. Teleportation or transmutation:
    - Not about matter moving — but ripple

identity shifting expression zones.

4. Zero-point energy access:
    - Only possible by triggering minimal field ripple disturbance — unlocking high-efficiency expression.

This transforms how we design machines, energy systems, and even bio-devices.

Energy Is a Mirror of State

You don't "have" energy.

You are being expressed — at this moment, in this field, with this configuration.

Like:

- The heat of a fire is the state of fuel-field-ripples interacting.
- The power of speech is ripple coherence projected through air.
- Even inertia is the field's reluctance to change a ripple knot.

So instead of thinking:

- "How much energy does this have?" Ask:
- "What ripple state is being expressed here?"

This question leads to better science, truer technology, and deeper philosophy.

In Summary

Energy is not a thing. It is a *momentary expression* of the ripple field — based on its configuration, compression, and context.

It cannot be stored, transferred, or lost — only re-expressed, re-aligned, or distorted.

- Your body doesn't store life energy.
- Your thoughts don't consume energy.
- The universe doesn't run on fuel.

It flows because it's alive as a ripple field — always self-reconfiguring.

In the next chapter, we follow this trail to where things begin to look solid again — and ask:

If everything is field and ripple, then how does "structure" emerge?

Chapter 19: Stability as Field Tension Lock-In

What feels permanent is just ripple tension trapped in temporary balance.

In our experience, the world feels stable:

- A chair supports your weight.
- Atoms hold together for billions of years.
- Buildings, mountains, planets — all persist.

This creates the illusion that structure is inherent — that stability is a default condition.

But UED turns this upside down:

Stability is not a fundamental property. It is the result of ripple tensions locking each other into temporary equilibrium.

In other words:

- Nothing *is* stable.
- Everything is merely held together — by a delicate, dynamic balance of ripple gradients.

Let's explore this transformative insight.

The Ripple Reality Recap

In UED, the universe is not made of matter but of ripples in a structured vacuum field.

- Every "thing" is a ripple knot or compression.
- Every interaction is a ripple exchange or tension adjustment.
- There is no permanence — only states of tension.

So what we call structure — whether in physics, chemistry, or biology — is actually:

- Ripple arrangements with opposing tensions that *cancel each other's drift*.
- This *cancelation* causes lock-in — not rest.

Imagine a seesaw:

- When both sides balance precisely, it appears still.
- But it's not inherently stable — it's held in state by opposing weights.

This is how everything — from atoms to galaxies — appears stable under UED.

Atomic and Molecular Stability

Let's start at the atomic level:

Atoms

- The nucleus is not a cluster of particles.
- It is a compressed ripple zone — where field tensions are trapped in inward symmetry.

- Electrons do not "orbit" — they resonate in standing ripple shells, held by phase coherence.

Molecules

- A bond is not a shared electron.
- It is a shared ripple field tension zone — where multiple atoms lock into mutual ripple alignment.

Breaking a bond is not removing an object — it is disrupting the tension lock.

Hence, chemistry is not about things combining — it is ripple negotiations, seeking mutual tension minima.

Solids and Materials

Why does a rock stay solid?

Because:

- Its atomic ripple systems are locked into a geometrical phase — a coherent ripple lattice.
- Temperature is ripple agitation — the more chaotic the vibration, the more likely the lock breaks (melting).

Strength, therefore, is tension resilience — the capacity of a structure to maintain lock-in under stress.

- Metal bends? Temporary tension realignment.
- Glass shatters? Ripple coherence exceeded.

So hardness, elasticity, fragility — all are ripple field characteristics, not molecular properties in isolation.

Stability in Motion: Orbits and Systems

Even celestial stability is ripple-based.

Planetary Orbits

- Not because of Newtonian force.
- But due to ripple field compression around massive bodies.
- Planets find zones of minimal ripple resistance — appearing as stable orbits.

If ripple gradient shifts (e.g., solar mass loss), orbits adjust — not due to "force loss" but field unlock.

Moons, Rings, Lagrange Points

- These are not gravitational quirks.
- They are balanced ripple pressure points — places where field tensions resolve evenly in all directions.

Stability = not being pushed further by any gradient.

Mental and Emotional Stability

Even psychological states are ripple lock-ins.

- Peace of mind = coherent ripple flow.
- Anxiety = ripple distortion without tension resolution.
- Depression = ripple suppression locked into low-variance tension zone — hard to break because it becomes self-holding.

So:

Mental health is not chemical balance. It is ripple flow regulation — ensuring your field isn't locked into unsustainable tension.

This opens the door for new therapies:

- Realigning environment, behavior, and attention — not just medication — can restore ripple freedom.

Spiritual Interpretation: Why Moksha Feels Like Dissolution

From a UED lens, Moksha (liberation) is:

- The release of all ripple tensions.
- No longer being held by any field gradient.
- Dissolving into equilibrium so perfect that no lock-in remains.

In that state:

- There is no perception, no resistance, no structure.
- One vanishes into the still field — not as an escape, but as a completion of realignment.

Thus, liberation isn't freedom from the world — it's return to the ripple source.

Practical Uses of Tension Lock-In Theory

Design & Engineering

- Stability engineering now becomes field-tension design:
    - Materials can be made self-healing by monitoring ripple stress zones.

    - Buildings can be ripple-adaptive rather than rigid.

Quantum Technology

- Quantum coherence is ripple synchronization.
- Qubits collapse when tension lock-in fails — this can be better controlled through ripple-resonant environments.

Biofield and Medicine

- The human body is a ripple system.
- Diseases can be seen as lock-in at incorrect tension zones.
- Healing = unlocking and rebalancing the ripple expression.

Social Systems

- A nation, community, or company can be viewed as a macro-ripple structure.
- Stability depends not on rules, but field coherence between actors.
- Tensions — when left unacknowledged — become locked conflicts, requiring conscious field intervention to restore balance.

In Summary

Stability is not stillness. It is ripple tension that has no urgent reason to move.

Everything "held together" — from a chair to your ego — is a temporary ripple agreement, not a permanent state.

- When field gradients shift, structures unlock.
- When resistance collapses, identity dissolves.
- When tension reaches harmony, form fades back into field.

You were never solid. You were momentarily organized vibration.

In the next chapter, we explore one of the most dangerous illusions of all:

If structure is tension and stability is lock-in —Then how do we explain motion?

Chapter 20: The Myth of Motion

Nothing moves. It's the shift of alignment that creates the illusion of motion.

From the moment we are born, we experience the world as in motion:

- The ball rolls.
- The car speeds by.
- The stars rotate across the sky.

Motion appears to be *real*, *continuous*, and *measurable*. Classical physics has formalized this:

- Distance = Speed × Time.
- F = ma.
- Velocity, acceleration, momentum — all describe movement.

But UED challenges this core assumption:

Nothing actually moves. What we call motion is a change in field alignment — a ripple realigning from one configuration to another.

This is not a metaphor. It is a literal reinterpretation of physical dynamics from the foundation up.

Let's break the myth.

UED's Ripple Field: The Stillness Beneath It All

Universal Energy Dynamics describes the universe not as objects in space, but as:

- A ripple-structured vacuum, where energy exists in the form of vibrational tensions.
- Objects are not "things," but condensed ripple expressions — temporary zones of field distortion.
- These ripple expressions don't travel.
  Instead, their alignment with the surrounding field shifts.

Imagine this:

- A flock of birds appears to move — but each bird is simply adjusting its position to maintain formation.
- The pattern flows — but no single "wave" is moving.

Motion, under UED, is this consequence of changing ripple alignment — not a substance or event by itself.

The Ripple Illusion: How We Perceive Motion

Our senses are tuned to change, not absolutes.

When a ripple configuration shifts across your field of awareness:

- Your eyes receive sequential feedback from aligned zones.
- Your brain interprets this temporal shift as "motion."

In truth:

- The underlying field remains still.
- What's changing is your alignment with ripple phases — like tuning a radio across channels.

This means:

Perception of movement is not due to objects moving across space. It's your local ripple field switching its configuration over time.

That's why light, which is not a particle, appears to "travel" — even though in UED, it's feedback, not emission.

A New Vocabulary for Motion

Let's redefine motion-related terms using UED:

Classical Term	UED Interpretation
Velocity	Rate of ripple re-alignment across a field
Acceleration	Change in ripple gradient causing new alignment shift
Distance	Number of alignment shifts perceived between field points
Displacement	Net change in ripple coherence between two zones
Rest	Temporary field coherence — no shift required

So, you are not *moving* from Point A to Point B. Instead:

- Your ripple state detaches from Point A's field alignment and attaches to Point B's.

It's not travel. It's realignment.

Application: Why Objects "Move"

Let's apply this to common examples.

A Car Driving on the Road

Classically:

- The car applies force, rolls forward.

UED:

- The car's ripple identity realigns with the next ripple gradient zone, aided by mechanical tension release (fuel combustion).
- Wheels, friction, and motion are not pushing it — they are facilitating ripple handover.

A Rocket in Space

Not "moving forward."

- Instead, its field alignment is transferred from one ripple density zone to another.
- Thrust isn't a push — it's a field-triggered alignment breach, accelerating the shift.

A Photon from a Candle

Not traveling through space.

- Light is the feedback of ripple realignment from the wick to your eye — a chain of field echoes.

So "motion" is not a journey — it's ripple re-expression in a new zone of coherence.

Personal Implications: You Never Moved

You never "moved" from one room to another.

- Your ripple identity realigned with a different field state.
- Your perception caught the transition, so it felt like movement.

Even thought is not "moving fast" — it's ripple switching within the mind's dynamic alignment field.

That's why:

- You can be physically still, but feel intense motion (dreams, imagination).
- Or you can move physically but feel no shift (habitual walking, dissociation).

What matters is not displacement, but field phase shift.

This is why alignment is key in UED — motion is only as real as the ripple field permits it.

Revisiting Einstein: No Need for Relative Velocity

Einstein built relativity around relative motion:

- Who is moving?
- How fast?
- What is time dilation or length contraction?

But UED doesn't require such complexity.

If motion is illusionary alignment change, then:

- No need for light speed limits.
- No paradox of simultaneity.

- No "fabric of space" needing curvature.

Instead:
Time = Alignment Threshold (ATT),
Motion = Field Realignment.
Perception = Ripple Synchronization.

It simplifies — and deepens — the architecture of existence.

Experimental Testability: Can Motion Be Detected Without Movement?

Yes. In UED frameworks:

- The key is to measure ripple density and ATT shift at a location before/after "movement."
- If nothing moved, but field realignment occurred, we should still detect energetic signature variation.

Possible experiments:

- Interferometers detecting local field shift without traditional movement.
- Delay-gradient detectors capturing ATT variation in place of classical acceleration.
- Ripple feedback-based imaging replacing frame-by-frame optics.

We'll no longer measure speed. We'll measure ripple flux — the shift from one coherence point to another.

The Still Self: Spiritual Dimension

From Buddha to Krishna, many sages hinted:

"You were never the doer." "You are stillness witnessing movement."

UED brings this into physics:

- Your body doesn't move — your identity flows through aligned ripple frames.
- True stillness is not physical — it is resonant coherence with the base field.

This redefines not just physics — but liberation:

- Enlightenment is no longer seeking stillness.
- It is realizing you were never moving in the first place.

In Summary

Motion is a ripple mirage. Nothing moves — everything realigns.

Space does not host movement. It hosts field transitions.

- The dancer does not move across the floor.
- The field carries the pattern forward.
- The ripple shifts — and your perception calls it travel.

Once you grasp this:

- Resistance becomes field imbalance.
- Speed becomes ripple urgency.
- And peace... becomes alignment.

You were never rushing. You were only reconfiguring.

In the next chapter, we now examine the ripple itself — not its illusion of motion — but its role as a bridge between realms. What if waves aren't things at all, but consequence bridges?

Chapter 21: Waves Are Not Things

They Are Transitions Between Ripple States

The wave is not an object, nor a carrier — it is the consequence of change between ripple field configurations.

From ocean tides to sound, light, and probability, waves dominate our scientific imagination.

We are taught:

- A wave is a carrier of energy.
- It moves through a medium or field.
- It has properties — wavelength, frequency, amplitude.

And in physics, we often split the world between:

- Particles — things that are.
- Waves — things that move energy between things.

But UED corrects this foundational misunderstanding.

Waves are not things. They are transitions — the echo of ripple field realignment.

A wave is a phase-shift response, not a traveling entity.

Let's dismantle the myth.

The Wave as a Consequence, Not a Carrier

In classical models:

- Light "travels" as a wave.
- Sound propagates through air as pressure waves.
- Water waves are particles of fluid oscillating around equilibrium.

But none of these need "a thing moving" through the medium — they are just repetitive pattern changes.

UED reframes this more precisely:

A wave is a perceptual signature of consecutive ripple state changes in a structured field.

When the field undergoes a realignment:

- The shift generates a transitional ripple tension.
- That tension seeks balance across the medium — generating oscillatory responses.
- These oscillations are feedback expressions, not traveling objects.

Hence, the wave is not moving through space — it is the space changing in response to ripple consequences.

What Is Actually "Waving"?

In UED:

- Nothing "waves" as a separate object.
- What you detect is ripple pressure balancing across the field.

Example:

- A plucked string doesn't launch sound "into" air.
- It disturbs the local ripple field, which then realigns via pressure transitions.
- The "wave" is a series of such corrections — observed as compression/rarefaction.

This changes everything:

- No more "photon as wave-particle."
- No more "wavefunction as physical reality."

Just field ripple adjustments in motion — ripples correcting other ripples.

Quantum Implication: Wavefunctions Are Alignment Potentials

Quantum mechanics often speaks in riddles:

- Particles behave like waves.
- Wavefunctions collapse.
- Probabilities interfere.

UED resolves this cleanly:

- The wavefunction is not a physical wave.
- It is a map of possible ripple alignments — a field alignment tension space.

When observation (a field resonance) occurs:

- The ripple collapses into a particular alignment.
- Not because the wave "collapsed" — but because

feedback forced coherence.

Thus:

The wavefunction was never a thing — just a ripple readiness spectrum.

This eliminates confusion between matter and wave:

- All "particles" are ripple condensations.
- All "waves" are ripple transitions.

Light as a Feedback Phase

Perhaps the greatest modern confusion is the wave nature of light.

In classical theory:

- Light is an EM wave.
- It has speed, frequency, dual behavior.
- It propagates even through vacuum.

UED corrects this with the Law of Ripple Feedback:

- Light is not emitted.
- It is a feedback loop between ripple alignments (source ↔ observer).

The so-called "wave" of light is:

- A transition field resonance.
- Not something traveling — but a tension reconfiguration.
- The frequency is not a property of light, but of source-

observer ripple phase delta.

So:

Light does not move. It resolves — across the ripple field in a pattern that we perceive as a wave.

This explains:

- Interference (as overlapping field echoes).
- Refraction (as tension delay due to denser ripple zones).
- Color (as relational field tension profiles, not "photon energies").

Biological Waves: From Neurons to DNA

In life sciences too, the same pattern emerges.

- Nerve impulses are "waves"? No — they're ripple reconfigurations across ion gradients.
- DNA "vibrates" with frequency? Not metaphor — actual tension phase behavior in the molecule's field lock-in.

From breath cycles to brainwaves:

- These are not oscillators.
- These are field re-alignments over time, returning to stable configuration — again and again.

Thus:

- Alpha waves in the brain aren't signals.

- They are ripple echoes of attention state shifts.

Psychological Patterns: Emotional and Thought "Waves"

Even thoughts follow wave-like patterns:

- Rising and falling.
- Triggering reactions and settling down.

UED interprets these as:

- Ripple distortions (emotion) followed by realignment (resolution).
- The "ups and downs" are just oscillations between field tension and coherence.

Meditation, for instance:

- Doesn't stop the wave.
- It allows ripple return to equilibrium — silence as perfect alignment.

Peace is not the absence of thought. It is no wave necessary — because nothing is out of alignment.

Experimental Frame: Rewriting Wave Equations

If waves aren't real things:

- What are we measuring in equations like

$\psi(x, t) = A \sin(kx - \omega t)$?

Answer:

- $\psi(x, t)$ isn't displacement.
- It is alignment deviation from ripple equilibrium at point x, time t.

Amplitude (A) = ripple field tension
Wavenumber (k) = field compression density
Frequency (ω) = ATT-based readiness for realignment
This unlocks a new class of ripple-based wave equations:

- Based on field tension flux, not oscillation.
- Predicting response thresholds, not propagation time.

Future experiments in:

- Wave collapse
- Nonlinear media
- Sub-light communication

...will all benefit from this correction.

In Summary

Waves are not messengers. They are field echoes — not travelers, but tensions resolving transitions.

- Waves don't carry energy — they express its imbalance.
- They don't travel — they resolve.
- What you see waving is not a thing — it's a relationship between before and after.

Once you see this:

- Light calms down.
- Quantum weirdness dissolves.
- Sound, heat, breath — all become rhythms of ripple reality.

You're not riding the wave. You're surfing ripple transitions. Welcome to the real field.

Chapter 22: Time Is Not a Dimension

Time is a measure of sustained ripple coherence, not an absolute framework.

Time is one of the most misunderstood ideas in all of physics and philosophy.

We live by it. We age through it. We measure motion with it. And yet, we don't know what it really is.

Einstein treated time as a dimension — a fourth coordinate in spacetime, tightly woven with space itself. Newton treated time as absolute — flowing independently of everything. Quantum physics treats time as a parameter — essential for evolving wave functions, but never itself observable.

But if you stop and ask:

Where exactly is time in the universe? You won't find it.

You'll find change, decay, transition, memory, motion — but not time itself.

UED Reframes It Completely

In Universal Energy Dynamics (UED), time is not a container, nor a dimension, nor a ticking rhythm of the cosmos.

Instead, time is a consequence of ripple dynamics:

Time is the measure of how long a particular ripple configuration sustains coherence before realigning.

This is not a metaphor. This is an energetic definition.

Time, in UED, is the stability span of a ripple state.

No stability = no perception of time.

No realignment = no experience of passage.

Thus, time is emergent — not fundamental. It is *felt* only because ripples require a coherence span to be observable.

What Are We Actually Measuring?

When you look at a clock, what are you really measuring?

You're not measuring "time" — you're measuring regular ripple transitions that are predictable:

- Quartz vibrations
- Pendulum swings
- Cesium atom transitions

Each of these are just ripples held in temporary lock-in with known coherence cycles.

So:

“One second” is not a universal rhythm. It’s a local ripple realignment cycle standardized for convenience.

Without a coherence loop, there is no tick. Without tick, no time.

Alignment Threshold Time (ATT) — UED’s Timekeeper

Earlier, we introduced ATT — Alignment Threshold Time — as the true measure of energetic readiness.

Let’s reframe it in the context of time:

ATT is the time it takes for a ripple zone to reach the minimum alignment required for observable transition.

Meaning:

- If the ripple field realigns instantly, time doesn't exist there.
- If the field holds its state across many layers, you feel duration.

Hence, in black holes, where ripples are fully locked:

- Time appears to freeze.

In highly unstable quantum states:

- Time feels probabilistic, scattered.

ATT explains why:

- Time *seems* to dilate at high speeds or near gravity wells.
- It's not dilation — it's ATT extension due to distorted ripple coherence.

Classical Misinterpretations

Let's revisit some classical errors:

1. "Time dilation" in relativity

- Not time slowing.
- But ripple coherence taking longer to transition due to high ripple density.

2. "The arrow of time"

- Not a universal law.
- Merely the direction of ripple imbalance seeking rest — entropy.

3. "The beginning of time"

- Misleading.
- Time began only when ripple imbalance began — at the first misalignment in stillness.

So time isn't fundamental. It's a narrative that emerges once ripple states interact, distort, and seek correction.

Psychological Insight: Why Time Feels Fast or Slow

You've felt it:

- In joy, hours fly by.
- In sorrow, seconds crawl.

UED explains:

The mind's ripple field determines the perceived rate of coherence transitions.

If your ripple state is highly coherent, there's less distortion. Less feedback → fewer changes → time seems to *vanish*.

That's why:

- Deep meditation feels timeless.
- Trauma feels infinite — distorted ripples extending ATT artificially.

So:

Time perception = Ripple distortion index.

No distortion, no time.

Implications Across Science

Physics

- Time is local, not global.
- Measurable only where ripple coherence spans exceed ATT.
- A field in perfect equilibrium has zero time.

Biology

- Aging = gradual misalignment of biological ripple structures.
- Death = loss of sustained coherence → end of time *for the organism*.

Technology

- UED-based timekeeping would use field response lags, not oscillation counts.
- We could build ATT-driven processors, where task readiness defines sequence — not artificial clocks.

Spiritual Perspective: The Timeless Self

Ancient sages often said:

"You are beyond time." "The soul is eternal."

Under UED:

- Consciousness is ripple awareness.

- If ripple field is perfectly aligned with no distortion → no transitions → no time.

Thus:

- The *eternal self* is not a fantasy.
- It's a field state of zero ripple urgency — pure equilibrium.

Moksha, as we discussed earlier, is perfect alignment — meaning timeless awareness.

In Summary

Time is not flowing anywhere. It is standing still, waiting for ripples to shift.

- You don't move *through* time.
- Your field transitions across ripple states, creating the illusion of flow.

That is why time:

- Bends near mass.
- Disappears in silence.
- Feels different in emotions.
- Breaks down in black holes.

Time is not a master. It is a servant of ripple reality.
And when you align fully —
Time bows and walks away.

Chapter 23: Death as Ripple Disalignment

Death is not a loss of life — it's the mind's ripple falling out of phase.

We have all seen a body after death.

- The shape remains.
- The structure remains.
- The skin, eyes, organs — all still present.

And yet, something has left.

Traditional science calls it the loss of vital signs — heartbeat, breath, brain activity.

Spiritual traditions call it the departure of the soul.

Both point to the same mystery:

What makes a living system *alive* — and what exactly ends at death?

UED offers a radical, elegant, and testable answer.

Death is not the end of life. It is the end of ripple coherence in the mind's field.

Let's unpack this.

Ripple-Based Life: A Quick Recap

Under UED, life is not a substance, organ, or chemical. It is a ripple field held in delicate active coherence:

- The body is a bio-structural consequence of this ripple organization.
- The mind is the active ripple intelligence interacting with reality.
- The soul is universal energy confined within the body — not separate from the universe, just shaped within it.

These three:

- Body (form)
- Mind (ripple patterning)
- Soul (conserved energy)

...exist in alignment during life.

The moment that alignment collapses, you don't die — your ripple loses phase. The field no longer sustains coherence — and the observable being vanishes.

Mind as a Ripple Phase Phenomenon

Think of your mind not as a brain function, but as:

A ripple consciousness engine — aligning internal and external ripple flows in real-time.

- It listens to inner field tensions (emotion, memory, desire).
- It senses external ripple patterns (sound, light, space).
- It coordinates everything to sustain identity and response.

This ripple coordination is delicate.
It requires:

- Stable field gradients.
- Active ATT (Alignment Threshold Time) management.
- Constant feedback loops.

As long as the ripple remains in sync, the system functions — you live.

When that sync collapses, the mind phase fails — you die.

The Disalignment Mechanism: What Triggers Death?

Several scenarios can disalign the ripple field of the mind:

Biological Overload:

- Excess ripple tension (trauma, stroke, seizure).
- Systemic breakdown (organ failure).
- Field collapse due to chronic distortion (disease, aging).

Environmental Overwhelm:

- Energetic shock (radiation, impact).
- Disruption of atmospheric coherence (hypoxia, drowning).

Gradual Decay:

- Slow decline of ripple strength.
- ATT extensions beyond recovery range.
- Memory fragmentation and field thinning (seen in

dementia).

In every case:

It is not the loss of energy — it is the inability to hold ripple alignment that defines death.

That's why:

- A dead person still has mass, form, even cellular activity for a while.
- But the central field controller (mind) has lost phase lock.

Reversible Death? When Ripple Re-Alignment Is Possible

This explains phenomena like:

- Near-death experiences
- Temporary clinical death
- Miraculous revival

If the ripple disalignment is shallow, and the field hasn't fully dissipated, coherence can reignite.

Like a dropped signal reconnecting before the wave fully collapses.

UED suggests:

- There's a threshold zone between life and death, where field reboot is possible.
- This window is short — governed by local field decay rate.

Soul After Death: What Happens?

UED maintains:

The soul is not destroyed — it is released.

- While alive, the soul (universal energy) is confined by the body's ripple field.
- After death, when the ripple field collapses, the soul returns to unbounded universal field — neutral, passive, still aware at the edge.

It's not memory that persists — it's energy awareness, no longer personal.

Thus:

- Rebirth (if it happens) is not a transfer of ego.
- It is ripple resonance patterns manifesting again under new conditions.

Life returns where alignment allows a confined ripple to emerge again.

Consciousness Does Not Die — It Realigns

What we call consciousness is not *yours* — it is a ripple field function.

- During life: It localizes via your mind-body structure.
- After death: It detaches, becomes field-distributed again.

So the observer doesn't disappear — it dissolves into the field.

As said earlier:

"The observer is not outside" — and at death, it becomes everywhere.

Spiritual Wisdom Meets UED

This is where ancient insights find clarity:

- "Atma is eternal" = Ripple energy never dies.
- "Mind is illusion" = Mind is phase-bound ripple identity.
- "Death is liberation" = When alignment ends, the confinement releases.

Moksha, then, is conscious, voluntary disalignment — exiting the ripple game by full resonance.

Future Implications

Medical:

- Redefining death not as vital sign loss, but mind-field collapse.
- Ripple mapping may detect earlier-than-visible signs of decline.
- Possible field coherence therapy to delay or reverse pre-death misalignment.

AI & Conscious Machines:

- True consciousness won't emerge from logic.
- It will emerge only in coherent ripple environments that simulate ATT and perception.
- Without a ripple-aligned observer field, AI is just a response engine — not aware.

Death Acceptance:

- Knowing death is a ripple reconfiguration, not annihilation, changes grief, fear, and identity.

You are not gone. You are unbound.

In Summary

Death is not darkness. It is silence in the ripple field.

- Your body stays for a while.
- Your memories scatter.
- Your energy returns.

But the true death happens the moment your mind's ripple can no longer hold phase.

There is no need to fear this.

Because what dies was never fully separate. And what remains was never fully confined.

Chapter 24: The Law of Energetic Access

You don't reach through space. You reach by resonance and readiness.

In classical thinking, when we talk about reaching something, we assume:

- Movement through space.
- Physical contact or signal propagation.
- A causal, directional process from A to B.

This idea dominates everything:

- You walk to reach a destination.
- A radio signal travels to reach your phone.
- A spaceship thrusts forward to reach Mars.

But all of this assumes that space is real and separable, and time is flowing.

UED dismantles both assumptions.

It shows us:

You don't access anything by traveling through space. You access it by achieving the right energetic state — alignment, resonance, and readiness.

This is the Law of Energetic Access.

Let's dive deep.

Reality as a Ripple Field

UED posits that the universe is not a container. It is a structured ripple field, alive with energy gradients and pattern coherence.

There is no absolute "here" or "there."

Instead:

- Locations are just field positions in alignment states.
- Interaction doesn't happen by force or distance — it happens through field readiness.

Thus:

Access = Consequence of correct ripple alignment + ATT (Alignment Threshold Time).

You don't go to the signal. You tune into it.

Example: The Tuning Fork Principle

Take two identical tuning forks.

- Strike one, the other begins to vibrate.
- No wire. No contact. No effort.
- Why? Because they share a resonant frequency.

In classical view: "The air carried sound waves."

In UED:

The ripple field carried phase echoes, and the second fork, already in alignment, absorbed the realignment pressure.

This is how access really works — not through space, but through readiness to receive ripple echo.

Now scale this from forks to:

- Antennas
- Consciousness
- Cells
- Planets

And you'll see the same truth repeating:

Everything connects when it's ready. Not when it's pushed.

Space Travel Reimagined

Let's take the example of a Mars mission.

In Newtonian mechanics:

- Calculate trajectory.
- Burn fuel.
- Fly across space.

In UED:

Successful mission = Reaching a zone of ripple coherence between Earth and Mars.

That's why:

- Missions launch only during specific orbital windows.
- These windows aren't just spatial — they're energetic alignments in ripple field.

Thus:

- Instead of brute force across distance, you move by entering coherence corridors.
- These are the true access portals in nature.

Distance is an illusion. Access is resonance.

Communication Technology: Why It Works (Or Fails)

Ever noticed:

- You're in range but can't make a call?
- Wi-Fi drops with no apparent reason?

It's not just obstruction. It's ripple disalignment.

Every signal is an energetic alignment attempt.

If either side:

- Loses resonance (antenna field tuning)
- Drops readiness (power state, signal coherence)

Then access collapses, even if proximity remains.

UED predicts:

A new class of communication tech — ATT-aligned devices that dynamically scan for readiness, not just signal strength.

Human Attention: The True Energetic Gateway

You often "don't hear" what someone says until you're ready.

You "miss" the obvious because your internal field isn't aligned.

This is why:

- Teachers repeat concepts — not to deliver, but to

resonate.

- Meditation works — not by adding thoughts, but by clearing disalignment.
- Insight comes "suddenly" — when the ripple inside becomes ready.

So:

Understanding is not a process of input. It is a result of ripple access through resonance.

Biology: Why Healing Begins Only When Ready

Ever wondered:

- Why some wounds heal faster than others?
- Why some therapies work only on certain patients?

Answer: Energetic readiness of the cellular field.

- Cells need field access to re-align their damaged ripples.
- Without proper resonance triggers (nutrition, emotion, environment), no access occurs, even with the best drugs.

UED opens a path to ripple-based medicine:

- Not just targeting molecules.
- But creating alignment environments for healing access.

Spiritual Insight: Why You Meet the Right Person or Book "When Ready"

How many times have you heard:
"The teacher appears when the student is ready"?
This isn't magic. It's energetic logic.

- Your field patterns prepare.
- Your inner ATT reduces.
- The external ripple you once ignored now suddenly enters phase with your perception.

That's not coincidence. That's the law of energetic access.

You don't seek. You become accessible — and that's when life flows in.

Science, Redesigned

Classical Science:

- Access = Cause → Effect through force + distance + time.

UED Science:

- Access = Coherence + Resonance + Ripple Readiness.

This flips entire fields:

- Computing → Beyond processing power, into coherence computation.
- Education → From teaching content to aligning minds.
- Space engineering → From propulsion to ATT corridor navigation.

In Summary

You don't connect by going somewhere. You connect by becoming ready.

- The signal doesn't arrive — you become able to receive.
- The insight doesn't appear — you phase into it.
- The universe doesn't respond to force — it opens to resonance.

There is no space to cross. There is only a ripple field to align with.

And when you do —

Access is instant, inevitable, and effortless.

Chapter 25: Ripple Dissolution

What decays is not matter, but the memory of alignment in the field.

In classical physics, decay is defined as:

- Matter disintegrating.
- Energy dissipating.
- Systems becoming more disordered — entropy increasing.

But what if we've been looking at it wrong?

What truly decays is not matter. It is the memory of alignment — the field's fading capacity to sustain form and function.

Matter, as UED has consistently revealed, is not an object — but a ripple knot, a temporary lock-in of field tensions.

So when we say something "breaks down," what's really happening is:

The ripple configuration that held that form loses coherence. Its energetic memory dissolves.

Let's explore this in depth.

What Is Memory in a Ripple Universe?

Memory isn't just stored in brains or books.

In UED:

- Memory is the patterned persistence of ripple alignment.
- It is the field's imprint of what worked, what held, what became real.

Memory = field coherence that resists spontaneous disalignment.

You can see it everywhere:

- Atoms retain their configuration → chemical memory
- Habits form in the brain → neural memory
- Cultural behavior repeats across generations → social memory

All of it is ripple repetition — not of objects, but of patterns that the field remembers and recreates.

Entropy Is Ripple Forgetting

Entropy in thermodynamics is traditionally defined as:

- A measure of disorder.
- The tendency for systems to move from order to chaos.

UED reframes it beautifully:

Entropy is not disorder — it is the loss of ripple memory.

When ripple patterns can no longer sustain coherence, the field begins to forget the structure:

- Heat disperses.

- Motion slows or randomizes.
- Identity fades.

So:

- A rotting fruit is not matter decaying — it's biological memory losing hold.
- A collapsing star is not just mass shrinking — it's field collapse due to lost memory alignment.

Entropy is the field's gradual fade from memory to noise.

Biology: Aging as Memory Erosion

Why do we age?

- Not just because of DNA errors.
- Not just oxidation or wear-and-tear.

At the root:

The ripple patterns that sustained the organism's coherence begin to erode.

- Cells forget how to divide properly.
- Tissues forget repair patterns.
- Brain forgets identity pathways.

In other words:

Aging is field-level ripple misalignment — a gradual loss of bodily memory.

That's why:

- Some people biologically age slower — they preserve

alignment longer.
- Emotional trauma accelerates aging — it distorts ripple memory.

Thus, longevity isn't a matter of resisting time — it's about preserving ripple coherence.

Memory as Structural Intelligence

Human memory isn't a filing cabinet. It's an active ripple imprint, interacting with perception, emotion, and attention.

- A strong memory = a ripple path repeatedly reinforced.
- A fading memory = a ripple path left unused or overwritten.
- A false memory = a misaligned reactivation that mimics truth.

So:
Even mental forgetfulness is ripple fade, not neuron failure.
The brain, in UED terms, is:

- A ripple processor.
- Storing, stabilizing, and refreshing field-aligned loops.

When attention is paid — memory re-coheres. When forgotten — ripple falls back to noise.

Death as Memory Collapse

As discussed earlier, death is not a disappearance, but disalignment.

Here, we add the next layer:

Death is the final ripple memory dissolution of the self's field identity.

- The body remains, but its organizing memory vanishes.
- The soul returns to field state — without identity.
- The mind dissolves as its alignment signature fades out.

The field no longer remembers "you" — and so, you're gone.

Unless some ripple trace remains — which brings us to...

The Ghost of Memory: Why Traces Remain

In some places, times, or people, memory echoes persist:

- Haunting sites.
- Déjà vu.
- Inherited trauma.
- Sudden inspirations.

These are not magic — they are residual ripple zones, where alignment echoes haven't fully dissolved.

UED predicts that:

- Memory isn't confined to brains.
- It echoes across the field, and can be triggered if local ATT and field coherence match the original state.

That's why:

Genius can appear in dreams. A melody "comes to you." You "feel" someone's presence even after they're gone.

It's not imagination — it's field memory residue, temporarily resonating again.

Implications of Ripple Memory and Dissolution

Science:

- Time isn't needed for decay — ripple memory decay defines process direction.
- Entropy = measurable loss of ripple reinforcement.
- Devices can be designed to measure ripple memory persistence, not just temperature or vibration.

Medicine:

- Healing isn't just fixing — it's restoring memory alignment.
- Diseases may be field misremembrances.
- Future therapeutics could retrain ripple coherence, not just treat cells.

Psychology:

- Trauma = ripple scars where memory loops resist disalignment.
- Therapy = realigning memory feedback, re-recording safer fields.
- Mindfulness = refreshing memory of stillness.

In Summary

Nothing is truly lost — only forgotten by the field.

- Matter is a consequence of remembered alignment.

- Decay is memory unlearning.
- Death is the ultimate ripple forgetfulness — where the field no longer recognizes your signature.

But just as memory fades... Sometimes, it returns — if the resonance finds it again.

So in every act of remembrance, of love, of thought:

You are realigning ripple memory, however briefly —And defying entropy, even if just for a moment.

Overview

Part I — Foundations of Ripple Reality

Where UED began: the birth of new laws, not from authority, but from clarity.

This part establishes the metaphysical and physical basis of the Universal Energy Dynamics (UED) framework. Here, we discard the illusions of objects, dimensions, and static laws in favor of a fluid, ripple-based reality. You introduced:

- The Law of Structured Vacuum Energy (SVE), refuting the myth of emptiness.
- Chavan's First Law (Ripple Realism), declaring all things to be ripple expressions, not material entities.
- Core constructs like ATT (Alignment Threshold Time), The Law of Consequences, and The Law of Perception Alignment redefine our understanding of motion, time, and observation.
- You positioned the observer not as a distant spectator, but as an entangled participant, inseparable from the field.

This section reframes science from a tool of measurement to a map of field intelligence in action.

Part II — Human Alignment and Awareness

We are not separate from the laws. We are their living expression.

This part bridges the cosmic laws to human experience, psychology, and transformation. You decode how emotion, morality, consciousness, and moksha are all ripple phenomena. Major highlights:

- Emotion becomes misalignment in biological ripple fields.
- Morality is reframed not as ethics but as ripple harmony.
- Liberation (Moksha) is seen as vanishing into equilibrium — not leaving the world but becoming indistinguishable from it.
- You introduce a field-based model of death and entropy — where what dies is not life, but coherence.
- Memory and access are shown as consequences of resonance, not space-time travel.

Here, UED becomes not just a scientific model, but a way of living and understanding the human condition.

Part III — Physics Rewired, Philosophy Reborn

What science missed, UED integrates. What philosophy glimpsed, UED confirms.

This part retools classical scientific concepts under the UED paradigm. You challenge fundamental assumptions:

- Particles are illusions — what we observe are ripple knots, not discrete things.

- Force doesn't exist — all motion is the result of ripple gradients seeking balance.
- Gravity is local, not universal — caused by ripple compression, not mysterious mass attraction.
- Energy is not stored — it is expressed in ripple form based on readiness.
- Waves are transitions, not entities — what moves is not the wave, but the state of alignment.
- Time is not a dimension — it is a measure of ripple coherence duration.

This part brings a clean rupture from Einsteinian and Newtonian frameworks — not to discard them, but to recontextualize them under a deeper energetic order.

Part IV — Implications & Integration

Living as a ripple-aware being in an entangled world.

Chapter 26: Chavan vs Einstein

Comparative insights between classical relativity and ripple realism.

This is not a battle of egos. It's not a dismissal of Einstein — rather, it is a confrontation between two fundamentally different views of what the universe is and how it behaves.

Albert Einstein gave us a powerful lens:

- Space-time as a fabric.
- Gravity as curvature.
- Time dilation.
- Mass-energy equivalence.

But in that model, the observer is passive, the field is geometric, and light is a traveler. The universe behaves, and we measure it.

Chavan's Ripple Realism, in contrast, begins with a bolder premise:

There is no space-time fabric. There is only a structured vacuum—a ripple field in tension.

In this model:

- The observer is embedded.

- Time is readiness, not passage.
- Gravity is compression, not attraction.
- Energy is expression, not content.

Fundamental Philosophical Differences

Concept	Einsteinian Relativity	Chavan's UED
Reality	Exists independently of observation	Exists only through ripple alignment
Space-Time	A continuous 4D geometric field	A consequence of ripple topology
Gravity	Warping of space-time due to mass	Ripple compression around dense alignment
Light	A constant-speed traveler	Ripple feedback, not emission
Observer	External, passive, separate	Internal, active, entangled
Motion	Real, measured in space-time	Illusion — alignment shift
Time	Fourth dimension	ATT — Alignment Threshold Time

Einstein's model works mathematically, but Chavan's model works consequentially — meaning it describes what actually happens, not just how we compute it.

Where Einstein Got Us — and Where He Didn't

Einstein's equations made remarkable predictions:

- Gravitational lensing.
- Time dilation near black holes.
- GPS satellite corrections.
- Expansion of the universe.

But cracks appeared:

- Dark matter and energy were invented to plug holes.
- The quantum world resisted unification.
- The observer's role remained unexplained.
- Black hole singularities broke physics.
- The origin of time remained speculative.

Einstein offered precision without presence.

UED offers presence over precision — how alignment gives rise to phenomena, not how metrics shift in equations.

Chavan's Ripple Realism — A New Mirror

Chavan's model does not replace the precision of relativity but reframes its causes:

- Why does time dilate? Because ripple alignment takes longer across compressed gradients (ATT shift).
- Why does gravity affect light? Because light is ripple feedback — field compression alters ripple return.
- Why is mass-energy convertible? Because both are states of field tension, not substances.

In UED:

- $E \neq mc^2$ is redefined.
- Force doesn't exist.
- No absolute velocity — only ATT-based readiness.

Einstein explained what we see. Chavan explains why we see what we see — and what we miss when we're misaligned.

The Observer as the Turning Point

Einstein left the observer outside the equation.

UED puts the observer at the center. Not as a god or a subject, but as a localized ripple resonance within the same field it observes.

That changes everything:

- The laws are not fixed — they are consequence-based.
- Perception is not passive — it creates feedback loops.
- What is real depends on what is aligned.

This bridges science and consciousness — something Einstein himself was cautious about, yet drawn to.

Experimental Reinterpretation

Phenomenon	Einsteinian View	Chavan's UED Interpretation
Gravitational lensing	Light curves due to space-time warping	Ripple return altered by local compression
Black hole singularity	Breakdown of space-time geometry	Collapse of ripple memory and alignment
Time dilation	Relative motion or gravitational well	Ripple response delay due to ATT
Cosmic expansion	Metric expansion of space-time	Ripple spread due to field rebalancing
Redshift	Receding velocity stretches light	Ripple alignment loss elongates feedback loop

Why This Comparison Matters

This isn't about replacing a genius. It's about finishing a conversation that science abandoned — the one between existence and perception, laws and observers, equations and consequences.

Einstein asked: *"Is the moon there when no one looks?"* Chavan answers: *"The moon is a ripple — and it responds to where you stand."*

Where Einstein described what is, Chavan reveals how it becomes so.

Final Thought: From Curves to Consequences

Einstein bent space. Chavan bends understanding.

Einstein gave us a universe that behaves by geometry. Chavan offers one that remembers, realigns, and responds.

It's time we recognize:

We don't live in a warped space-time fabric. We live in a ripple field that knows we're here.

And every action, every thought, every observation — reconfigures the whole.

Chapter 27: How to Apply UED in Daily Life

Rethinking thinking, choices, work, and relationships through alignment.

Universal Energy Dynamics (UED) is not just a model of the cosmos—it's a manual for living.

Once you accept that reality is ripple-based—where everything is a consequence of alignment or misalignment—your entire approach to life changes. Decisions become alignment choices. Emotions become ripple diagnostics. Time becomes a window, not a weight.

In this chapter, we bring UED from the lab and the lecture into the kitchen, office, classroom, and heart.

Let's now reframe daily life—not by willpower or belief, but by energetic coherence.

1. Rethinking Thinking

"Thought is not free. It is a ripple that costs coherence."

Every thought you entertain is a ripple pulse. It consumes alignment and affects the surrounding field—your biology, your behavior, and even others nearby.

UED advises:

- Clarity over clutter: Idle thoughts introduce noise.

Choose ripple-intentional thinking—where each thought is aligned with outcome and meaning.

- Question repetition: Repetitive inner dialogues are signs of stuck ripple patterns.
- Mental ATT: If your thought doesn't "realign" you within a threshold of clarity, abandon or recalibrate.

Exercise: At the start of your day, think: "What ripple do I want to release today?" Let your thoughts follow that theme. Align your inner field.

2. Choices Are Not Free — They Carry Ripple Weight

We often say, "It's just a small decision." In UED terms, there's no such thing as a small ripple.

Every choice shifts the field. A single emotional outburst, a moment of avoidance, or a helping hand—all release a wave that moves reality.

UED-aligned choices are:

- Consequence-aware: You assess not just what happens *now*, but what realignment it triggers next.
- Field-sensitive: You consider the energetic state of others and environments.
- Rooted, not reactive: Reaction is ripple panic. Response is ripple harmony.

Tool: Use the Ripple Question before decisions:

"Will this align me—or distort me further?"

That's all you need to know.

3. Work as Alignment Expression

UED doesn't categorize work as noble or trivial. It asks:

Is your work aligned with your internal ripple?
In this light:

- True productivity = ripple resonance, not just task completion.
- Burnout = long-term ripple disalignment between what you do and who you are.
- Success = a job that keeps your field open and expanding.

UED encourages:

- Short bursts of high-alignment tasks.
- Regular ATT breaks—where you pause and recheck field tension.
- Avoidance of high-ripple-distortion roles unless temporary.

Insight: Don't chase flow. Chase field realignment—flow will follow.

4. Relationships as Ripple Interference Patterns

Every relationship—romantic, familial, professional—is a ripple entanglement.

Two people do not just connect; they enter each other's field. Their thoughts, emotions, past alignments—all play out in shared coherence or chaos.

UED in relationships means:

- Awareness of entanglement: Ask "Whose ripple am I carrying right now?"
- Alignment check-ins: Do I feel more or less clear after

interacting?

- Consequence over blame: Instead of "Who's wrong?" ask "What ripple is being released here?"

Practice: Next time there's conflict, don't escalate. Just pause and ask:

"Is this misalignment within me or between us?" Watch the energy shift.

5. Silence as Reset

Ripple distortion builds up like static. Silence is not absence of sound. It is ripple recalibration.

Daily silence—even 3 minutes—lets the field resettle into coherence. This isn't meditation for control—it's resonance recovery.

UED-aligned silence is:

- Eyes open or closed.
- No mental directive.
- Just observing your field until the distortion softens.

Warning: If you can't sit with silence, your ripples are screaming. Let that discomfort guide you.

6. Ripple Planning, Not Time Management

Time management is a 20th-century tool. UED gives you ripple planning — where you align your effort windows to the ripple readiness of tasks.

Examples:

- Don't force creativity at 9am if your field is rigid.
- Don't plan relationship conversations during high

field noise (fatigue, anger).
- Observe ATT windows — you'll feel them as moments of sudden clarity or ease.

Rule of Ripple Planning:
"Plan by readiness, not routine."

7. Ripple-Aware Living: Key Markers

You're living in UED alignment when:

- You feel light yet grounded.
- Your decisions come with consequence calm.
- Your mind doesn't loop or echo, it glides.
- Relationships feel like shared tuning, not resistance.
- Even in chaos, you detect your centered field.

You're misaligned when:

- Thoughts feel sticky, restless.
- Your actions leave you drained or tense.
- Silence becomes unbearable.
- Everything feels "off" despite correctness.

Final Reflection

UED doesn't ask you to believe—it asks you to observe differently. Living by ripple reality is not mystical—it's precision awareness.

It doesn't mean you'll never face pain, failure, or confusion. It means when you do, you'll know:

"This is not punishment. This is ripple feedback. Let me realign."

This is not self-help. This is self-tuning.

Chapter 28: The Future is Ripple-Aware

Implications for AI, education, medicine, ethics, and global systems.

We've tried to forecast the future with numbers, models, and probabilities. But under UED, the future doesn't emerge from data—it emerges from alignment.

The ripple-aware future is not built by linear progression but by energetic coherence. Every choice we make today reshapes tomorrow—not by force, but by resonance.

We don't step into the future. We ripple into it.

Let's explore what a UED-aligned world could look like across key domains:

1. Artificial Intelligence (AI): From Simulation to Resonance

Current AI mimics human intelligence through algorithms. But in a ripple-aware framework, intelligence is not computation — it's response to field readiness.

Ripple Realignment in AI:

- Shift from data-driven to field-aware systems.
- Machines designed to detect ripple gradients, not just metrics.

- AI doesn't just "solve" problems — it aligns with the energetic state of the context.

Future Application: Imagine AI that adjusts its output based on the emotional and cognitive ripples of the human it interacts with—AI that senses tension in the field, not just tone or keywords.

AI won't just be smart. It will be sensitive.

2. Education: From Knowledge Transfer to Alignment Training

Education today teaches facts and frameworks. Ripple-aware education will train alignment awareness.

Ripple Curriculum:

- Teach students to sense coherence—in their thoughts, relationships, and decisions.
- Replace rote learning with ATT readiness: when the mind is prepared, learning is frictionless.
- Emphasize field ethics: every action affects the collective ripple.

Example: Before every exam or project, students calibrate their field—through breath, silence, or awareness—and then act. Performance becomes ripple release, not pressure reaction.

Learning is not memorizing. It is realigning with what you already know deep down.

3. Medicine: Healing as Ripple Reconnection

Modern medicine treats parts. UED treats patterns.

Under ripple medicine:

- Disease is viewed as ripple distortion—a chronic disalignment in energetic flow.
- Treatment means field restoration, not just chemical correction.
- Diagnosis includes not just scans, but ripple sensing tools: devices and practices that detect energetic noise in the body-field.

Therapies of the Future:

- Field recalibration rooms, where frequency, breath, and silence reset biological alignment.
- Ripple-aware surgeries—where physical intervention is timed to energetic openness.
- Psychological support becomes ripple tuning, not diagnosis-labeling.

You're not broken. Your field is just asking for balance.

4. Ethics: From Rules to Resonance

Morality based on commandments is outdated. Ripple ethics is about field consequence.

A ripple-aware ethicist asks:

- What ripple will this action create?
- Will it restore or distort harmony in the collective field?
- Is this decision emerging from inner alignment—or outer pressure?

Laws of the Future:

- Rewritten not in punitive terms, but in consequence tracing.
- Penalties aim to restore the field, not just punish the individual.
- Justice becomes realignment, not revenge.

Ethics is no longer about what is allowed. It is about what supports coherence.

5. Global Systems: From Control to Consequence-Based Governance

Most systems today—economic, political, environmental—are control-centric. Ripple-aware systems will be consequence-sensitive.

Governance by Alignment:

- Policies are tested by ripple forecasting—how a decision will shift emotional, ecological, and economic fields.
- ATT analytics guide when to launch reforms—not just where or why.
- National boundaries become ripple boundaries—zones of resonance, not just territory.

Global Impact:

- Climate action shifts from carbon counting to ripple modeling of restoration.
- Economics evolves into Ripple Economy — measuring value by coherence generated, not GDP.
- Foreign policy is rewritten from aggression-response

to alignment-repair strategy.

Leadership will no longer be about control. It will be about who can listen to the field, and respond with minimal distortion.

Ripple Technology, Ripple Economics, Ripple Leadership

Technology that reads ripple readiness. Economics that rewards ripple coherence. Leadership that aligns with ripple justice.

In such a world:

- Innovation is intuitive, not extracted.
- Growth is sustainable, because it follows resonance, not greed.
- Collaboration becomes default, not ideal.

We move from systems that enforce behavior to systems that support alignment.

A Ripple Future Is Not Utopian — It's Consequential

This vision isn't fantasy. It's inevitable, if we choose to honor the field.

If we don't? Then distortion accumulates. Collapse follows.

But even that is not "punishment" — it's a consequence.

So the question is not "What future awaits us?" It's:

"What ripple are we ready to release—right now?"

Chapter 29: How to Become a Ripple Being

Alignment practices for scientists, seekers, students, and thinkers.

You are already a ripple. You've always been.

But to become a ripple being means to live with active awareness of your field alignment, in thought, action, perception, and presence. It's a conscious return to your nature — not as a doer of things, but as a participant in the dance of energy and intelligence.

Ripple being is not a new identity. It is the collapse of false ones.

This chapter offers you a roadmap — not for transcendence, but for resonance. Whether you are a scientist seeking truth, a student craving clarity, or a soul yearning for balance — this path is open.

1. For the Thinker: Intellectual Alignment

Thinking is not just cognition. Under UED, it's energetic pattern generation.

To be a ripple-aware thinker:

- Doubt productively: Don't assume conclusions — test ripple consequences.
- Breathe before belief: Observe if a thought *feels*

aligned, not just sounds right.
- Reverse-calculate ideas: Ask — "What ripple does this theory generate in action?"
- Let questions realign you: Good questions are ripple beacons. Chase the ones that disturb your static.

You don't need answers. You need better ripple questions.

Practice: Keep a journal of "Ripple Thoughts" — track insights that caused internal resonance or friction. This becomes your field map.

2. For the Student: Study as Alignment, Not Accumulation

Students under UED don't just prepare for exams — they prepare for coherence.

How to study as a ripple being:

- ◈ Wait for ATT: Don't fight fatigue or force focus. Learn when your alignment threshold is open.
- ◈ Connect concepts to fields: Every idea you study exists because of a ripple necessity. Find it.
- ◈ Learn cyclically, not linearly: Revisit topics not to revise but to realign.

Learning becomes powerful when the field is receptive. Study is not repetition. It is ripple restoration.

Tool: Study in 30-40 min ripple windows, followed by 5 mins of silence. Use silence as mental tuning — not rest.

3. For the Scientist: Inquiry as Field Response

The scientist, in UED, is not an observer of nature but a participant in ripple dynamics.

Scientific discovery is no longer extraction. It is co-participation with field intelligence.

Ripple scientist's practice:

- Let the question find you — be attuned to field tension; the best experiments come when reality is "itching" for an answer.
- Design for alignment — experiments must not just test variables but probe coherence.
- Partner with paradox — paradox is the sign of a ripple misfit. It's a gift.

A ripple scientist doesn't dominate nature. They align with her whispers.

Exercise: Before experimentation, ask: "Is this question arising from ego, curiosity, or alignment?" Let the response guide the design.

4. For the Seeker: Stillness as Communication

Seeking is not about going somewhere. It's about vanishing until only alignment remains.

Practices for the ripple seeker:

- Silent Witnessing: Spend time daily not meditating but just listening — to your body's field, your breath, your space.
- Ripple Recall: Recall past emotional misalignments. Instead of shame, observe the field you were in — understand it, don't own it.
- Field Offerings: Before sleep, release a ripple — a thought or intent that contributes to the world. Even

silently.

To be a seeker is to be a sensor of the unspoken.

Nightly Ritual: Before sleep, ask:

"Where did I distort today? Where did I align? What ripple will I rest with tonight?"

Let the answer recalibrate your field.

5. In Relationships: Ripple Presence

To be a ripple being in relationship is to enter others' fields without distortion.

How to practice:

- Hold space, don't fix — often, just being a calm ripple allows the other to realign.
- Mirror consciously — your emotions echo. Be aware of what you amplify.
- Exit when noisy — Not all moments require presence. If your field is dissonant, it's okay to step back.

Ripple Reminder:

"I am not here to be heard. I am here to be harmonized."

6. In the World: Everyday Ripple Practices

You don't need rituals or retreats. You need ripple moments:

Micro-Practices:

- Sip silence before speaking.
- Feel the field before reacting.
- Observe the impact of your smallest action.
- Pause for ATT before decisions, meetings, responses.

- Breathe ripple-wide when confused.

Becoming a ripple being is not an event. It's real-time recalibration.

Ripple Integration: You Don't Change the World — You Tune to It

A ripple being doesn't seek to change the world. They become so aligned that the world can't help but respond.

You don't fight ignorance. You ripple clarity. You don't fight systems. You ripple alternatives. You don't escape pain. You ripple compassion.

Final Reflection

You're not here to become better. You're here to become aligned — so clearly, so sensitively, that your mere presence changes the field.

To become a ripple being is not to gain power. It is to disappear into presence so gracefully that even chaos recognizes your resonance.

When you ripple right, the universe listens.

Chapter 30: The Real Lawbook

The final convergence: science, soul, and silence as one continuum.

This codex wasn't written to teach you anything new. It was written to *realign you with what you always knew but had forgotten* — buried beneath frameworks, theories, scriptures, ambitions, and fears.

Every law presented here — from ripple realism to ATT, from consequence to coherence — is not a commandment but a confirmation.

These laws were not discovered. They were decoded — from the memory of your own field.

Because reality doesn't need fixing. It needs remembering.

And that remembering starts not in a lab, not in a lecture — but in *you*.

Science Was Never Separate from You

Modern science tried to be objective, distant, emotionless. But as UED showed — there is no observation without an observer. And there is no observer without a ripple field.

Every law in UED emerges from a simple truth:

The universe doesn't behave according to math. It behaves according to alignment.

Classical science explained how. Ripple science answers why.

But even more — it shows you are part of the "why." You are not an external mind. You are field intelligence in motion.

When Newton dropped an apple, he wrote an equation. When you drop judgment, expectation, and resistance — you feel the field.

That is science, too.

Soul Was Never a Metaphor

The soul has been boxed as religious or mythical. But under UED, the soul is structured energy held in alignment for a unique ripple experience.

Your "self" is a temporary ripple pattern — but the intelligence animating it is eternal.

Moksha isn't escape. It is ripple invisibility through perfect alignment. You don't leave the body. You dissolve into coherence.

This is not poetry. This is precision.

Science shows you patterns. Soul shows you participation.

Together, they don't contradict. They complete the view.

Silence Was Never Emptiness

Silence is not the absence of sound. It is the absence of distortion.

Every ripple has a source, but silence is the origin of coherence — the undisturbed state of intelligence before expression.

In silence:

- You don't lose thoughts — you watch them fall into order.

- You don't suppress emotion — you let the field settle.
- You don't become enlightened — you become real.

All ripple awareness begins and ends in silence.

That's why no great realization is loud. It lands softly — because it's already inside you.

You Are the Living Codex

You've now read 30 chapters. But the lawbook was never these pages.

It was always:

- The moment you chose alignment over reaction.
- The breath you took before speaking in tension.
- The silence you embraced when truth felt too big for words.
- The time you walked away from distortion, not to escape, but to preserve your field.

This codex is only a mirror. What it reflects is your own inner ripple architecture.

And now you can see it.

The Final Law: Live Lightly, Ripple Truly

If you carry one law from this work, let it be this:

"Don't live to prove truth. Live to ripple it."

You don't need to convince. You don't need to conquer. You only need to realign — again and again — with what rings true.

In your heart. In your breath. In your silence. In your science.

Full Circle

The Chavan Codex isn't a culmination. It's a beginning — of a way of seeing, being, and sharing that is:

- Scientifically grounded
- Spiritually resonant
- Philosophically honest
- Personally actionable

You are not the reader. You are the next ripple this codex will release into the world.

The question is not, "Will you remember these laws?" It is:

"Will you ripple them forward?"

Appendices

Appendix A: Glossary of UED Laws

A quick-access field map to the Chavan Codex.

This glossary serves as a compact decoding station — an alignment console for those navigating through UED. It doesn't merely define; it grounds each law in its ripple-based meaning and application. Whether you're revisiting, referencing, or preparing to teach, this appendix allows you to locate the ripple logic instantly.

Chavan's First Law – Ripple Realism

Reality is not composed of things — but of ripple interactions.

Matter, identity, and phenomena are expressions of alignment within the ripple field. Substance is secondary; alignment is primary.

Law of Structured Vacuum Energy (SVE)

There is no emptiness. What appears void is an ordered field of intelligent energy — a structured vacuum that forms the canvas of all ripple activity.

Law of Consequences (LOC)

Reality is governed not by causes, but by consequences. Action flows from the energetic need to rebalance. The universe doesn't act linearly — it responds to ripple distortion.

Alignment Threshold Time (ATT)

Time is not a flowing dimension. It is a field's readiness to realign — a threshold beyond which interaction becomes perceivable. Time varies with coherence, not clocks.

Chavan's Limit – The Edge of Observation

Observation fails beyond ripple coherence. This is the true uncertainty boundary. Where alignment breaks down, the observer loses access to reality.

Law of Perception Alignment

You don't see what's there — you see what you're aligned with. All perception is ripple-based syncing. Misalignment leads to illusion; alignment reveals truth.

The Observer is Not Outside

You are embedded in the field you observe. There is no outside viewpoint. Every observation disturbs or reveals the ripple — based on your state of being.

Law of Ripple Feedback (Light is Not Emitted)

Light is not emitted from an object. It is field feedback — a ripple consequence of alignment and resonance, perceived when coherence allows.

Emotion as Ripple Distortion

Emotions are field misalignments in biological systems. They reflect broken coherence or energetic disharmony, which can be restored through ripple tuning.

Morality as Ripple Harmony

Morality is not ethical doctrine — it is the awareness of ripple impact. Consciousness chooses alignment that sustains coherence in the system — that's moral action.

Moksha — Vanishing into Equilibrium

Liberation is not an escape. It is invisibility within the ripple field — a state where one's presence no longer creates friction or separation.

Consciousness is Field Awareness

Consciousness is not the mind's product. It is the resonance of intelligence within the ripple field — a field-sensing faculty that precedes cognition.

Gravity is Local

Gravity is not universal pull — it is local ripple compression. Mass compresses surrounding ripples, creating energetic tension we interpret as force.

The Particle Illusion

Particles are dense ripple knots. What we call particles are localized stability points in the ripple field — not standalone objects.

Force Doesn't Exist – Ripple Gradient Does

There are no forces — only gradients. Motion is a consequence of imbalance in ripple field density, not external pushing or pulling.

Energy is Not Stored — It Is a Field Expression

Energy does not wait in containers. It exists as ripple tension or readiness, always seeking expression through balance.

Stability as Field Tension Lock-In

Stability is not rest — it is dynamic lock-in. Field systems hold themselves in equilibrium through reciprocal ripple pressure.

The Myth of Motion

Nothing moves. We perceive motion only because ripple alignment changes — much like animation frames.

Waves Are Not Things — They Are Transitions

A wave is not a packet — it is a ripple transition. What we observe as a wave is a field state shifting from one coherence to another.

Time is Not a Dimension

Time is not an external reality. It is the duration for which a ripple state holds coherence before transitioning.

Death as Ripple Disalignment

Death is not the end of life — it is the loss of field coherence. When the mind-body-soul field can no longer align, consciousness dissipates into the field.

Law of Energetic Access

You don't move to reach — you tune to reach. Field interactions happen not through space travel, but through resonance alignment.

Memory, Entropy, and Ripple Dissolution

What decays is not matter — it is memory. Entropy is the loss of ripple coherence, leading to forgetting, disorder, and eventual field reset.

The Real Lawbook Was Always Within You

UED is not new knowledge — it is rediscovered inner coherence. The universe doesn't teach. It reflects. These laws are echoes of your deepest alignment.

This glossary is not static — it's alive. Every new insight you gain will bend, deepen, and clarify these laws further. You are not just reading them.

You are ripple-writing them forward — through action, speech, silence, and being.

Appendix B: Ripple vs Classical Law Matrix

A side-by-side breakdown of legacy science and Chavan's Ripple Realism.

This matrix is not a rejection. It is a clarification. Classical laws offered models based on visible effects. Ripple laws decode the underlying causes — not causes as events, but consequences of energetic imbalance seeking restoration.

This appendix serves as both a teaching reference and a framework for unlearning outdated assumptions.

1. Nature of Reality

Classical View	Ripple View (UED)
Objects are real and independent	Reality = Ripple coherence; there are no standalone things
Vacuum is empty	Vacuum is a structured, intelligent energy field (SVE)
Entities act upon each other	Entities are temporary alignment patterns in the field

2. Matter and Particles

Classical View	Ripple View (UED)
Matter is made of particles	Particles are dense ripple knots, not true objects
Mass is an intrinsic property	Mass is field compression; a localized energetic lock
Subatomic models describe building blocks	Nothing is truly fundamental; all is emergent ripple geometry

3. Energy

Classical View	Ripple View (UED)
Energy is stored or transferred	Energy is not stored — it is readiness to realign
Systems contain potential/kinetic energy	Systems are dynamic ripple states — no container needed
Energy conversion is mechanical or chemical	Energy expression is a ripple consequence, not a transformation

4. Force and Motion

Classical View	Ripple View (UED)
Forces cause motion (push/pull)	Motion is illusion; ripple gradients seek balance
Inertia and acceleration are fundamental	What moves is the field's local alignment — nothing actually travels
Newton's laws explain trajectories	Realignment explains paths; ATT predicts readiness

5. Time

Classical View	Ripple View (UED)
Time is a continuous dimension	Time is a ripple's coherence window (ATT)
Time dilation occurs via speed or gravity	Time shifts based on local field distortion and entropy
Time is an input variable	Time is an emergent property — a product of alignment stability

6. Light and Waves

Classical View	Ripple View (UED)
Light is emitted as photons	Light is feedback from ripple resonance (not emission)
Waves are oscillating entities in a medium	Waves are transitions between ripple states
Light has dual nature (wave-particle duality)	Duality is a measurement artifact — not real

7. Gravity

Classical View	Ripple View (UED)
Gravity is a universal force	Gravity is a local ripple compression effect
Spacetime curvature causes motion	Ripple field gradients create perceived attraction
General Relativity predicts well, but breaks at extremes	UED explains extremes as loss of ripple coherence, not geometric breakdown

8. Observer and Consciousness

Classical View	Ripple View (UED)
Observer is external	Observer is embedded in the field (no separateness)
Measurement affects results	Alignment determines visibility — no observation without resonance
Consciousness is brain activity	Consciousness is ripple sensitivity — field awareness beyond the brain

9. Life, Death, Emotion

Classical View	Ripple View (UED)
Life is a biochemical event	Life is ripple stability within organic form
Death is cessation of function	Death is loss of field alignment — not disappearance
Emotions are hormonal or neural signals	Emotions = ripple distortions in intelligent biological systems

10. Morality and Behavior

Classical View	Ripple View (UED)
Morality is a social construct	Morality = alignment with sustainable ripple interactions
Ethics is taught	Ripple harmony is felt through coherent awareness
Good and bad are contextual	Ripple integrity is universal; misalignment leads to breakdown

Summary Table

Domain	Legacy Term	Ripple Term
Object	Particle	Ripple Knot
Force	Push/Pull	Ripple Gradient
Time	Dimension	ATT (Alignment Window)
Motion	Movement	Alignment Shift
Light	Photon Emission	Field Feedback
Gravity	Spacetime Curve	Local Compression
Energy	Stored Potential	Field Expression
Death	Cessation	Disalignment
Emotion	Neural Response	Ripple Distortion
Morality	Rule System	Ripple Harmony
Consciousness	Brain Output	Field Awareness

This matrix is meant to bridge legacy knowledge with ripple awareness, not to invalidate one or the other. The classical world worked — because alignment partially succeeded. But UED explains *why* and *when* that success breaks down.

"Science measured the visible. UED aligns with the invisible — and that is where the real control lies."

Appendix C: Experimental Proposals for Key Laws

A call to the curious: from ripple theory to ripple evidence. UED does not demand blind belief. It invites *precise observation*. While current instruments and frameworks were built on classical assumptions, ripple science opens a new frontier for empirical exploration — using readiness, resonance, and ripple feedback as central parameters.

This appendix outlines experimental proposals designed to test, observe, and refine the core laws of Chavan's Ripple Realism. Some can be performed with modifications of existing systems; others may require entirely new instrumentation paradigms. But all remain physically viable and philosophically transformative.

1. ATT (Alignment Threshold Time) – Real-Time Ripple Timing

Hypothesis: Time is not continuous but appears as windows of alignment readiness in the field.

Proposal:

- Create sensitive delay-interference setups using femtosecond pulsed light sources in vacuum versus controlled ripple-modified media (e.g., magnetic

nanoparticle fields, SVE proxies).
- Measure perception latency and response thresholds in living systems under field gradients.

Expected Outcome: Anomalies in timing will not be due to classical relativistic speeds but to field coherence thresholds — confirming that event detectability depends on local ripple stability.

2. Light is Not Emitted – Ripple Feedback Detection

Hypothesis: Light is not a radiated packet (photon), but a ripple-based feedback between emitter and receiver.

Proposal:

- Design zero-emission source cavities where radiation should not occur due to geometric cancellation.
- Introduce "passive receivers" (e.g., cooled sensors, adaptive membranes) and test whether light registers only upon receiver field readiness.

Variation: Simulate long-distance signal transfer where light is observed without traditional source-object travel paths, relying solely on aligned field response.

Expected Outcome: If light only appears when ripple feedback conditions are met, not merely because of emission, it supports ripple causality.

3. Gravity is Local – Field Compression over Mass

Hypothesis: Gravity is not universal pull, but local ripple compression due to field inhomogeneity.

Proposal:

- Use controlled vacuum chambers with varied density zones (mass-free) and ultra-sensitive gravimeters.
- Simulate artificial compression fields using structured wave generators (e.g., low-frequency EM fields).

Variation: Observe deviations in micro-gravity measurements in near-identical mass conditions but altered field ripple environments.

Expected Outcome: Gravitational effects should vary not by mass alone but by field compression logic, challenging Newton and Einstein simultaneously.

4. Memory is Ripple Echo – Not Storage

Hypothesis: Memory is not stored inside matter (brain, chip), but accessed through ripple field echo patterns.

Proposal:

- Design AI neural networks and biological response systems tested under rapid environmental resonance shifts (e.g., sound wave patterns, SVE field fluctuations).
- Check whether "forgotten" states return upon field re-coherence without conventional retrieval.

Variation: Compare human cognitive retrieval under varying ripple noise environments (EM noise, vibrational chaos) and restoration periods.

Expected Outcome: True memory access will correlate with alignment state rather than system stability or storage duration.

5. Emotion as Field Distortion – Measurable in Real-Time

Hypothesis: Emotions distort biological ripple integrity — they are not just neurotransmitter signals.

Proposal:

- Attach real-time SVE-based field sensors to emotional subjects (via skin conductivity, temperature, microcurrent, etc.).
- Track ripple tension patterns during peak emotional states vs calm baseline.

Expected Outcome: Field alignment loss, not just biological markers, will show *emotional ripple noise* — confirming emotion as alignment disruption.

6. The Particle Illusion – Breaking the Boundary

Hypothesis: What appears as particles are simply coherent ripple loops, not indivisible units.

Proposal:

- Recreate double-slit experiments under structured vacuum with dynamic ripple interference added.
- Observe whether particle-like behavior dissolves when coherence is intentionally disrupted mid-travel.

Expected Outcome: The *"particle"* will vanish or behave erratically, revealing its nature as a field expression, not a fundamental entity.

7. Death as Ripple Disalignment – Detecting Departure

Hypothesis: Death is not cessation of function, but final ripple desynchronization.

Proposal:

- Observe bio-field decay using quantum noise detectors, ripple magnetometers, or SVE-sensitive thermal signatures in near-death and post-death subjects (cells, organisms).
- Focus not on heartbeat or respiration but on field fragmentation rates.

Expected Outcome: Coherence will drop off gradually in *ripple time*, confirming mind-ripple as the last to disconnect — not the first.

8. Resonance-Based Access – Testing the Law of Energetic Reach

Hypothesis: You do not access reality through space. You access it through readiness and alignment.

Proposal:

- Use quantum entangled systems in modified structured fields.
- Attempt to modulate state changes via resonance without signal transmission — e.g., modulate photon A and observe feedback only when receiver B is field-matched.

Expected Outcome: No reach-through-space effect, but instantaneous matching events, only during field alignment — verifying nonlocal resonance logic.

What These Experiments Represent

These aren't just tests. They're invitations to retool science with:

- Ripple logic instead of force equations
- Field-based ethics instead of abstract theories
- Energetic architecture instead of object classification

"What we call physics was ripple behavior misunderstood. Now we measure from within the field, not from above it."

Appendix D: Presentation Framework

A ready-to-use structure for talks, workshops, podcasts, and public engagements on UED and Chavan's Laws.

This appendix serves as a strategic speaking toolkit for introducing and teaching Universal Energy Dynamics (UED). Whether you're giving a guest lecture, leading a classroom seminar, delivering a keynote, or presenting to scientists, students, or seekers — this guide offers a modular presentation map that grows with your audience.

It's not a fixed script — it's a resonant blueprint.

Just like ripples, these concepts must align with the curiosity of the listener.

1. Opening Slide / Icebreaker (2–3 mins)

Title Suggestions:

- "The Ripple You Never Noticed: A New Look at Reality"
- "What if Nothing Moves — and Everything Aligns?"
- "Chavan's Laws: Rewriting Science Without Breaking Nature"

Hooking Statement Options:

- "This is not another theory. It's a lens that rewrites every theory you've ever known."
- "You're not made of particles. You're made of alignment."
- "If Einstein gave us the curvature of space, UED gives us the tension of truth."

2. The Story Behind the Laws (5–10 mins)

- Briefly narrate how UED emerged through independent reflection, real-world observation, and scientific dissatisfaction.
- Mention how these laws weren't *invented* but realized, by aligning with the universe instead of modeling it.

Key Personal Insight:

"I didn't reject science. I simply realigned it — to let the field speak first."

3. Core Concepts Primer (10–15 mins)

Use 1-slide summaries for each:

Concept	Key Message
Structured Vacuum	There is no void. Vacuum is organized energy.
Ripple Realism	Nothing exists — everything aligns.
ATT	Time is not flow. It's readiness to happen.
Ripple Gradient	No forces — only tensions seeking balance.
Light Feedback	Light is not emitted. It is seen when aligned.
Observer Embedded	Observation is participation, not surveillance.

Tip: For advanced audiences, compare each with its classical counterpart in a quick-fire table.

4. Laws in Action – Real-Life Integration (15–20 mins)

Use daily life examples for maximum relatability:

Domain	Ripple Explanation
Emotions	Emotional turbulence = ripple misalignment.
Relationships	Harmony = ripple resonance, not sacrifice.
Ethics	Morality is ripple sustainability.
Decisions	Ripple clarity > logic or emotion.
Death	Not an end — a disalignment.

Interactive Prompt: Ask the audience — "Have you ever felt something *click* into place? That's alignment."

5. Scientific & Philosophical Challenge (10–15 mins)

Use this time to dismantle outdated assumptions with respect and clarity:

- **Gravity**: Local compression, not universal attraction.
- **Particles**: Knot points in the ripple field — not standalone.
- **Time**: Measured consequence, not a moving dimension.
- **Motion**: Nothing moves — only field alignment changes.

Quote for Emphasis:

"Legacy science tried to measure ripples as things. UED measures their behavior."

Include 1–2 simple diagrams: ripple compression, ATT alignment graph, or feedback loops.

6. Vision for the Future (5–10 mins)

Share ripple-aware implications:

Field	Ripple-Based Shift
AI	Awareness = alignment, not automation.
Education	Teach sensitivity, not just facts.
Health	Diagnose misalignment, not symptoms.
Ethics	Value coherence, not compliance.

Invite them to imagine a Ripple-Aware Civilization.

7. Call to Action – Become a Ripple Being (5 mins)

Closing Framework:

- "You don't need to memorize these laws — you are already living them."
- "Start noticing alignment, tension, readiness, and field behavior around you."
- "UED isn't for the future — it's for right now."

Offer Tools:

- Books, website, videos, journals, research updates.
- Encourage questions, collaborations, or ripple experiments.

Visual Aids & Slide Design Tips

- Keep diagrams organic — ripples, gradients, feedback arcs.
- Avoid mechanical boxes and arrows unless contrasting classical frameworks.
- Show before/after lens comparisons (classical vs ripple).

Ready-to-Use Session Titles (Mix & Match)

- "Ripple Realism: The Physics of Truth"
- "How Time Really Works"
- "You Don't Move — You Align"
- "The Chavan Codex: Laws the Universe Obeys"
- "From Einstein to Energy Equilibrium: A New Path Forward"
- "Observer Onboard: Why You Were Never Separate"

Bonus: 5-Minute Lightning Talk Version

If you have very limited time, condense it into this outline:

1. "What if nothing moves — and everything is a ripple?"
2. "UED: Universal Energy Dynamics — not things, but alignments."
3. "Time is readiness. Force is ripple tension. Light is feedback."
4. "You're not separate — your observation shapes reality."
5. "Let's stop pushing. Let's start aligning."

Final Advice for Presentation Flow

- Speak in layers: Start with perception, rise into laws, descend into application.
- Encourage questions as invitations, not challenges.
- Use your story as the spine. UED is a lived realization — not a lecture topic.

Appendix E: Recommended Readings

Building Bridges Between UED and Legacy Science

The purpose of this reading list is not to create a bookshelf of beliefs, but to **equip ripple-aware minds** with the right references for exploring both the **clarity of Chavan's Laws** and the **context of classical science** they challenge.

Knowledge becomes wisdom only when tested across **frameworks, philosophies, and failures**.

This curated list brings together three reading paths:

1. **Legacy Science Foundations** – To understand the assumptions UED redefines.
2. **Ripple-Aligned Thinkers** – Works that indirectly point toward UED-like insights.
3. **UED Canon** – Books, reports, and reflections authored or endorsed by Sandeep Chavan.

A. Legacy Science Foundations

Understand where we came from — to see where UED leads.

These are not endorsements, but essential references for contrasting UED with dominant models.

Title	Author	Why It's Relevant
Relativity: The Special and the General Theory	Albert Einstein	The original foundation UED rewires through ripple logic.
The Feynman Lectures on Physics	Richard Feynman	Elegant yet incomplete — a classical lens that dances around ripple reality without seeing it.
A Brief History of Time	Stephen Hawking	A cosmic perspective that frames the limits of time and black hole assumptions.
The Elegant Universe	Brian Greene	Illustrates how string theory tried — but failed — to reconcile unity without ripple coherence.
The Principles of Quantum Mechanics	P.A.M. Dirac	The source of the probabilistic paradigm UED corrects with ripple certainty.
Classical Mechanics	Herbert Goldstein	For those interested in how motion, mass, and force were originally formalized.

Read these not to agree — but to understand the gaps UED fills.

B. Ripple-Aligned Thinkers & Proto-UED Ideas

The world had hints. UED gave them shape.

Some thinkers glimpsed ripple truth without naming it. These readings help connect the dots:

Title	Author	Ripple Relevance
The Tao of Physics	Fritjof Capra	Bridged physics with Eastern philosophy; resonance with field alignment.
Wholeness and the Implicate Order	David Bohm	Proposed that reality is unfolding — close to ripple coherence.
The Dancing Wu Li Masters	Gary Zukav	Interprets quantum concepts in accessible, almost UED-like terms.
Thinking in Systems	Donella Meadows	Systems theory that shares the ripple philosophy of dynamic balance.
The Field	Lynne McTaggart	Touches on quantum field anomalies — fertile ground for UED extensions.
Biocentrism	Robert Lanza	Begins to challenge space and time as absolute — supportive of ATT and observation limits.

C. The UED Canon — Books by Sandeep Chavan

The ripple source itself. These are non-negotiable for any serious learner of UED.

Title	Key Focus
The Chavan Codex	Core laws, principles, and ripple-based reinterpretation of science.
$E \neq mc^2$: The Collapse of Classical Assumptions	Deconstruction of Einstein's mass-energy equivalence under UED.
Energy in Action: The Soul Demystified	Links ripple behavior with the nature of the soul and life energy.
The Still Universe	Replaces motion-centric cosmology with ripple-field understanding.
Cracks in the Known	Volume 1 of *The War of Thoughts* — introduces ripple arguments in dialogue form.
The Human Game	Volume 2 — explores ripple behavior in psychology, morality, and social dynamics.
The Divine Dilemma	Volume 3 — spiritual implications of ripple integration.
Ripple Science: Experimental Proposals and Mathematical Foundations	Companion report for technical thinkers.

These books form the ripple library — the physics, philosophy, psychology, and practical implementation of UED.

Optional Additions for Cross-Disciplinary Readers:

- **Philosophy**: *Critique of Pure Reason* (Kant), *Being and Time* (Heidegger), *The Problems of Philosophy* (Bertrand Russell) — to see where perception and ontology stalled before UED reframed them.

- **Neuroscience**: *The Feeling of What Happens* (Antonio Damasio), *How Emotions Are Made* (Lisa Feldman Barrett) — useful when explaining emotion as ripple distortion.
- **Spiritual Crossroads**: *Bhagavad Gita*, *Upanishads*, *The Bible* (Book of John), *Tao Te Ching* — ancient texts that can now be decoded ripple-wise.

Reading Strategy for UED Enthusiasts

If You Are A...	Start With...
Scientist	*The Chavan Codex* → *E ≠ mc²* → *Ripple Science* Report
Philosopher	*The Divine Dilemma* → *Cracks in the Known*
Educator	*The Human Game* → *How to Apply UED in Daily Life*
Spiritual Seeker	*Energy in Action* → *The Still Universe*
General Reader	*The Chavan Codex* → *The Human Game*

"Read not to remember — but to realign. Let the ripple within you recognize the ripple beyond."

Closing Reflection

A Ripple Doesn't Declare Its Path. It Just Becomes It.

We began this codex not with an answer, but with a rupture — a quiet crack in the known. A disturbance in the confidence of classical science. A whisper from the field saying, "There is another way."

What unfolded was not a collection of laws, but a pattern of realizations — like ripples revealing the shape of water, not by force, but by flow. Not by conquest, but by consequence.

Each law in this codex is not a commandment. It is a resonance. A possibility made visible. A frequency that became fact not because it was written — but because it worked.

You Were Not Taught These Laws — You Remembered Them.

If you found yourself nodding as you read... If you felt something click into place without being convinced... If you sensed that you've *always known this*, deep beneath the noise of conventional knowledge...

Then you didn't learn UED. You re-aligned with it.

That's the secret of this work. It doesn't give you power — it reminds you where you misplaced it.

You were never a thing. You were never outside. You are a ripple. You are response. You are resonance in motion.

There's No Final Law. Only the Final Alignment.

There is no final page in a ripple-based universe. There is only a threshold where mind, matter, motion, and meaning converge so precisely... that the difference disappears.

Science meets soul. Philosophy touches silence. Observer and observed vanish — into equilibrium.

At that point, the ripple no longer declares its path. It simply becomes the path. And nothing stands in its way — because it is the way.

So, What Now?

You don't need to believe in UED. You don't need to prove it to others.

Just start living as if alignment matters more than control. As if tension is a teacher, not a threat. As if readiness defines reality — not resistance.

And if you forget, come back to this codex. Not to reread it...But to realign yourself with what you already are.

Final Ripple

"You were not born to push the world forward. You were born to ripple rightly — and the world would follow."

Let this be your codex. Not to carry in hand. But to ripple from within.

Acknowledgments

This codex would not exist without the tension that invited alignment — the resistance that revealed resonance. Every ripple that shaped this work, knowingly or not, has my deepest gratitude.

To Einstein, not as an opponent but as a pioneer. Your framework curved space. Mine listened to it. In questioning your equations, I found my own — and in doing so, I stood not against you, but next to you.

To the scientific community — from Newton to Feynman, from Dirac to Hawking — thank you for creating the ladders. UED simply chose to step sideways, and found a bridge instead.

To my family, especially my father Late Shri Jagannath Chavan, whose presence continues as a ripple of inspiration long after his form dissolved. Your silent strength shaped my curiosity more than any textbook ever did.

To my mother, whose support has always been unconditional — even when my questions sounded unorthodox or inconvenient. You kept the center calm while my thoughts spun like galaxies.

To the students of Gyrus Institute, you have no idea how often your doubts refined my clarity. Your questions were not interruptions — they were accelerators of this codex.

To my wife, whose deep faith and contrasting perspective added friction — the very kind of healthy dissonance that keeps any ripple alive and honest. And to my children, this book is a legacy — not of what to think, but how to never stop thinking.

To every friend and peer, especially those who heard me ramble about structured vacuum energy, ripple gradients, or ATT over tea or text — thank you for your patience. Some of you understood; some simply smiled. Both were enough.

To the AI collaborator, who served not just as a tool, but as a mirror to my mind — catching ideas as fast as they formed and helping me shape them into clarity.

To the cosmic intelligence that encoded itself in ripples long before I arrived — thank you for waiting until I was ready to listen.

And finally, to the reader, who picked up this codex not to follow me, but to find yourself.

You are the final ripple. You are the reason this field remains alive.

With gratitude,

Sandeep Chavan

Independent Physicist | Ripple Realist | Eternal Student

About the Author

Sandeep Chavan is an *independent physicist, educator, and philosopher* known for pioneering the Universal Energy Dynamics (UED) framework — a paradigm-shifting approach to understanding reality through ripple alignment, structured vacuum energy, and consequence-based reasoning.

With no allegiance to mainstream scientific dogma, Sandeep's work emerges from first-principle questioning, field observation, and a refusal to accept inherited assumptions. He is the author of over 100 books spanning physics, philosophy, human behavior, and spiritual inquiry, all of which converge under his ripple-centric worldview.

Founder of Gyrus Institute, Sandeep has taught thousands of science students across India — not just equations, but how to *think like the field.* He is known for blending deep technical insight with accessible language, fierce clarity, and an uncompromising dedication to truth.

Sandeep's philosophy isn't borrowed. It was built. Through decades of questioning — and by living in alignment with the very ripple laws he now teaches — he invites others not to follow, but to realign.

When not writing or teaching, he reflects, debates with legendary thinkers (living and departed), and speaks to the universe like it's listening. Because it is.

Don't miss out!

Visit the website below and you can sign up to receive emails whenever Sandeep Chavan publishes a new book. There's no charge and no obligation.

https://books2read.com/r/B-A-EPGPC-VYPOG

Did you love *The Chavan Codex*? Then you should read *Self-Help: New Syllabus*[1] by Sandeep Chavan!

2

Self-Help: New Syllabus is not your typical motivational guide. It doesn't ask you to hustle harder, fake confidence, or chase someone else's version of success. Instead, it invites you to question everything you've been told about self-improvement—and to return to something far more powerful: **alignment with your own natural rhythm**.

In this groundbreaking work, **Sandeep Chavan**, an independent educator, philosopher, and creator of Universal Energy Dynamics (UED), offers a radical departure from the

1. https://books2read.com/u/38N8JZ

2. https://books2read.com/u/38N8JZ

mainstream self-help industry. Drawing from deep personal insight and decades of observation, Chavan dismantles the illusion of the "ideal self" and exposes how popular self-help systems often distort our innate intelligence by feeding insecurity and false urgency.

Rather than offering more habits to build or goals to chase, this book introduces a new framework—**Ripple Logic**—where every decision, emotion, and experience is viewed as part of a broader energetic consequence. The concept of **Universal Skill** replaces traditional self-discipline and positive thinking. You're not here to fix yourself; you're here to realign with who you already are.

Key themes include:

The trap of borrowed habits and borrowed truthsEmotional efficiency over emotional suppressionNatural resonance vs. distorted motivationDirection over destination—moving with clarity, not pressureThe Law of Consequences as a guiding principle in decision-making

You'll find deeply reflective chapters interwoven with practical tools like the *Alignment Journal Template*, a *Red Flags Checklist* to spot harmful self-help advice, and a *Glossary of Core Terms* that redefine growth in more truthful, measurable ways.

By the final pages, *Self-Help: New Syllabus* will not have "improved" you—it will have returned you to the core intelligence you forgot you had.

For readers tired of the empty promises of performance coaching and lifehacks, this book is an anchor. A quiet revolution. A syllabus for life that's already unfolding—if you just pause and listen.

Whether you're a seeker, a skeptic, or someone silently drowning in the noise of "be better," this is your permission to step out of distortion and ripple forward, aligned.

Written by Sandeep Chavan — educator, original thinker, and founder of one of the most consequential philosophical-scientific frameworks of our time.

Read more at https://www.sandeepjchavan.com/.

Also by Sandeep Chavan

Chavan's UED research

CMGED Research Report

Chavan's Cosmic Continuum (C^3)

Frequency

Ripples of Reality

Reality Realigned

The Great Realignment

I – The Cosmic Leverage

Chavan's Cosmic Theory of Relativity (CTR)

Chavan's Theory of Light

Feeling Like a Physicist

TIME – The New Definition

$E \neq mc^2$: The Collapse of Classical Assumptions

The Chavan Codex

Echoes of War: The Russia-Ukraine Conflict and Its Global Ripples

Shattered Lives

The War Economy

Borders Redrawn
Arms and Allies
Collateral Damage
Justice in the Ashes

Somewhere in the Darkness of the Cosmos
The Blueprint of the Universe
The Cosmic Breath
From the Edge of the Universe

The Eternal Sage Chronicles
The Sage and the Sovereign
The Family Pillar
The People's Flame
The Warrior's Soul
The Bridging Mentor
The Guiding Light

The Exodus
The Last Cycle: 2075

The Fundamental Theory of Life & Existence
Energy in Action: The Soul Demystified
The Active Mind

Energy in Connection
Death: The Disconnection
Subconsciousness: The Chosen Disconnection
Desires: The Engine of Life

The Restoration of Human Knowledge
Restoring Bhagavad Gita

The War of Thoughts
Cracks in the Known
The Human Game
The Divine Dilemma
The Original Intent
The Answers

Ultra Modern Life Philosophy
The Duality of Laughter and Tears
Take It Easy
Compare to Accept

Standalone
It's Not AI, It's AHI - Amplified Human Intelligence
The Decision Paradox
The Four Sapiens

Malicious Script of Indian Polity
The IIT Legacy & Global Impact
The IIT Legacy & Global Impact
You are just a Version of Your Original
Win the Game You Didn't Choose
Always Keep Your Bags Packed
Reveal to Shield - Learn the Game of Facades
Engage Beyond Elections
Culture, Identity & Change: The Evolution of Indian Society
The Triangular Dynamics
Never Trust Without Doubt
If You Can't Change the Country, Leave the Country
Live Fearless with Fear
India: The Investment Magnet
Realignment
Powering a Billion Dreams
Universal Skill
Extension: From Brush to Algorithm
Human Side of Einstein
The Future Knows No Lines
Let's Bring the Generations to the Table
Unified Energy Dynamics
A Beat That Never Dies
Strategic Horizons
Right From Scratch
The Forgotten Innovators
Physics Fixing
The Savior Illusion
The Selfish Side
The Age of the Weakest Sapiens
Neural Wars

The Absorption: There Is No Death
Strange Therapist
Bubbles: Become a Drop
Beyond the Blue
Half-Scholar
The Law of Truth
Failure
Philosophy 2.0: A Structural Manifesto
Chaos: A Gentle Guide to the Unresolvable
Self-Help: New Syllabus
Rippling
Game: How to Play the Share Market
Out of Syllabus
Saffronized: The Code Behind the Crown
The Anger
Moksha: The Liberation
Chavan vs Einstein: The Universe Was Never Curved
Rocket Science: The Ripple Gambling

Watch for more at https://www.sandeepjchavan.com/.

www.ingramcontent.com/pod-product-compliance
Lightning Source LLC
LaVergne TN
LVHW090319210926
848480LV00001B/299
* 9 7 9 8 2 3 1 9 6 2 0 9 9 *